C000083771

About The Author

Ian Wishart is an award-winning journalist and author, with a 30 year career in radio, television and magazines, a #1 talk radio show and five #1 bestselling books to his credit. Together with his wife Heidi, they edit and publish the news magazine *Investigate* and the news website www.investigatedaily.com.

Other recent titles:
Vitamin D: Is This The Miracle Vitamin? 2012
The Great Divide, 2012
Missing Pieces, 2012
Daylight Robbery, 2011
Breaking Silence, 2011
Arthur Allan Thomas: The Inside Story, 2010
Air Con: The Seriously Inconvenient Truth, 2009

For Heidi and all my children

TOTALITARIA

Ian Wishart

HOWLING AT THE MOON PUBLISHING LTD

First edition published 2013
Howling At The Moon Publishing Ltd
PO Box 188, Kaukapakapa
Auckland 0843, NEW ZEALAND

www.howlingatthemoon.com
email: editorial@investigatemagazine.com
Copyright © Ian Wishart, 2013
Copyright © Howling At The Moon Publishing Ltd, 2013

The moral rights of the author have been asserted.
Totalitaria is copyright. Except for the purpose of fair reviewing, no part of the publication may be copied, reproduced or transmitted in any form, or by any means, including via technology either already in existence or developed subsequent to publication, without the express written permission of the publisher and copyright holders. All Rights Reserved.
ISBN 978-0-9876573-5-0

Typeset in Adobe Garamond Pro and Frutiger
Cover concept: Ian Wishart, Heidi Wishart, Bozidar Jokanovic
Book Design: Bozidar Jokanovic
Front cover photo: NEWSCOM

To get another copy of this book airmailed to you anywhere in the world, or to purchase a fully text-searchable digital edition, visit our website:
WWW.IANWISHART.COM

LEGAL NOTICE: Criticisms of individuals in this book reflect the author's honest opinion, for reasons outlined in the text or generally known at the time of writing

Contents

If You Have Nothing To Hide...

Neo: 'What is the Matrix?'
Trinity: 'The answer is out there, Neo, and it's looking
for you, and it will find you if you want it to.'
– The Matrix

If you follow that hoary old chestnut to its oft-quoted ending, you would not fear any of what you are about to read in this book. Unfortunately, it is one of the most cunning propaganda phrases ever unleashed into the conversation of ordinary or garden variety members of the public. It is designed to imply that only the criminally guilty have any reason to hide their affairs from lawful authority. In the gaze of public and media pressure, who amongst us would not drop their trousers at the request of authority and expose our buttocks to the nice Homeland Security officer at the airport?

That's exactly what happened to American-born Shoshana Hebshi after a flight from California to Detroit:

"During her several hours in detention, Ms. Hebshi was subjected to an invasive and humiliating strip search, which required her to strip naked, bend over, and cough."[1]

It's a more modern take on the leery old doctors' motto, "please remove your top and say 'ahhh'."

---

1 http://www.aclu.org/files/assets/01_-_complaint.pdf

Hebshi's crime, although American by birth, was to be of Middle Eastern ethnicity and seated next to a couple of likely characters from South Asia on the flight who were acting suspiciously and both went to the toilet for long periods. Hebshi didn't know the men, she was a victim of random airline seat allocation. In fact, had they bothered to ask they would have found Shoshana was the symbol of multicultural-ism – daughter of a Jewish mother and Saudi Arabian father, married to a doctor, and a freelance journalist to boot. For all we know, perhaps that last fact was her real crime.

Alternatively, perhaps the security officials were suspicious that anyone would voluntarily fly to Detroit, a city now bankrupt and in large portions turned to ruin, where even cattle and chickens can be found foraging on abandoned city blocks.[2]

Regardless, Shoshana was dragged off the plane at gunpoint, as her lawsuit against the federal government lodged 2013 makes clear:

"Agents and employees of the FBI, the TSA, CBP, and Wayne County Airport Authority Police collaborated and put into place a plan to divert and board the aircraft, arrest Ms. Hebshi and the two men sitting next to her, and remove them to a detention facility at the airport for questioning.

"At approximately 4:25 p.m., Defendants Carmona, Bohn, Johnson and Defendant ICE Special Agent Brumley, along with other officers, boarded the plane, heavily armed, and ran down the aisle where Ms. Hebshi and the other two men in her row were seated.

"Several officers shouted at the passengers to keep their heads down and put their hands on the seat in front of them. Ms. Hebshi was stunned when the officers stopped at her row and yelled at her and the two men seated beside her to get up."

Shoshana soon found herself in an interrogation room with an officer named Toya Parker, and was instructed to drop her tweeds for Uncle Sam:

"Defendant Parker took off Ms. Hebshi's handcuffs and told her to remove all clothing, including her underwear and bra, so that she was completely naked. Defendant Parker instructed Ms. Hebshi to stand

---

2 "Can't solve the problems because the problems are cultural," Washington Post columnist George Will told ABC. "You have a city, 139 square miles. You can graze cattle in vast portions of it. Dangerous herds of feral dogs roam in there. You have 3 percent of fourth graders reading at the national math standards. Forty-seven percent of Detroit residents are functionally illiterate. Seventy-nine percent of Detroit children are born to unmarried mothers. They don't have a fiscal problem, Steve. They have a cultural collapse." http://dailycaller.com/2013/07/28/george-will-detroit-doesnt-have-a-fiscal-problem-but-a-cultural-collapse

facing the wall, away from the video camera, so that at least part of her body would be concealed.

"Ms. Hebshi was instructed to bend over, spread her buttocks, and cough while Officer Parker stood a couple of feet away and watched. Defendant Parker then instructed Ms. Hebshi to take her hair down from its ponytail so Defendant Parker could feel through Ms. Hebshi's hair. Defendant Parker lifted Ms. Hebshi's eyelids and looked in her mouth."

Lifted her eyelids?

The long and short of the story is that despite having every orifice searched, Shoshana Hebshi ended up having nothing to hide. It also turned out she had everything to fear.

"Before Ms. Hebshi was returned to her cell, her handcuffs were removed, and she was fingerprinted and asked her date and place of birth, weight, and height. Defendant Unknown TSA Officer 2 then came into Ms. Hebshi's cell with Ms. Hebshi's phone and required that Ms. Hebshi show the Twitter messages she had sent out from the airplane upon landing, as well as her Facebook profile.

"At approximately 7:30 p.m., Defendant Brand authorized the release of all three suspects. Ms. Hebshi was finally allowed to call her husband and let him know that she was okay and could leave soon. As soon as she started to speak to her husband, Ms. Hebshi cried."

One can understand, post 9/11, people and flight crews being jumpy about suspicious looking Middle Eastern passengers. But surely a handshake and a discussion over a coffee in the interrogation room is a more productive introduction than a booty-shake and a slap on the rump.

After that experience, you might be thinking, 'who in their right mind would book a flight to Detroit these days?', but on a more serious note Shoshana's story encapsulates the risk of assuming that just because you are innocent, your day won't turn to custard at the hands of the Government.

I began this chapter with the infamous quote, if you have nothing to hide you have nothing to fear. A more accurate restatement of the proposition is:

"If you have reason to fear, you have reason to hide."

The essential question this book asks is a simple one: do you have reason to fear? Have we reached a point where the power of the State, worldwide, has become overbearing and unreasonable? Have we given our respective bureaucracies too much power in the name of creating a protective nanny state?

This is not a book about shadowy conspiracy theories. Everything in this book happened, as is described. The details are meticulously footnoted and cited, where possible with a web address as well so you can check it for yourself.

As you will discover, the agenda is actually "hidden in plain sight" for the most part and there's nothing shadowy about it. What this book aims to do is bring together different parts of the jigsaw and reassemble them to make it easier to see the big picture.

Only then can you truly make an informed decision as to whether the powers of governments, aided by modern technology, have tipped the balance too far in favour of the State and its lobbyists.

This, also, is where the "slippery slope" argument kicks in. Prick your ears up and sit up straight, because based on the evidence now emerging we are on the brink of a powershift tipping point, or a point of no return.

The danger is that, for the first time ever, the State is morphing into a global entity which will be too large for any individual, or indeed any individual nation, to challenge. It is, in a sense, a political arms race in which the outcome is a kind of 'superstate' capable of legislating and enforcing to protect its own existence, even against the protests of the people.

If you don't take my word for it about the implications, then perhaps you will listen to the words of David Rockefeller, the 98 year old patriarch of his family who, perhaps more than anyone else over the past century, has done more to push globalisation as the endgame:

"Global interdependence is not a poetic fantasy but a concrete reality that this century's revolutions in technology, communications and geopolitics have made irreversible. The free flow of investment capital, goods and people across borders will remain the fundamental factor in world economic growth and in the strengthening of democratic institutions everywhere," says Rockefeller in his 2002 *Memoirs*.

Americans watching hundreds of thousands of illegal aliens crossing their borders every year have more reason than most to wonder about Rockefeller's vision, but Europe and Australasia are equally destinations for immigrants both legal and illegal. The removal of national borders was, ironically, a key foundation for the founder of communism, Karl Marx, as well because, like Rockefeller, he knew it would create the social mood for a new world.[3]

---

3 "On the Question of Free Trade" by Karl Marx 1848, http://www.marxists.org/archive/marx/works/1848/01/09ft.htm#marx

"In general, the protective system of our day is conservative, while the free trade system is destructive. It breaks up old nationalities and pushes the antagonism of the proletariat and the bourgeoisie to the extreme point. In a word, the free trade system hastens the social revolution. It is in this revolutionary sense alone, gentlemen, that I vote in favour of free trade."

With free trade eventually comes freedom of movement, although it is not yet fully operational as US theologian Daniel Groody noted in a paper on migration:

"The fact that in our current global economy it is easier for a coffee bean to cross borders than those who cultivate it raises serious questions about how our economy is structured and ordered."[4]

However, these men – including Rockefeller for all his sins – are only saying exactly what the premise of this book is: that new technology and political aspirations have made a single command post for the entire planet not only possible, but almost unavoidable. For the first time in human history those with ambition who nurse a genuine desire to control the world might actually achieve it. The questions you should be asking are no longer 'if' this will happen but 'when', and who are those ambitious people? What do they stand for? What do they believe?

To one extent, the title of this chapter is correct: if you don't buck the status quo, if you accept the limitations on freedoms your ancestors fought hard to gain, if you don't stand out from the crowd and just go about your own business, doing what you are told, then you probably do having nothing to fear. But if, one day, you decide the State has gone too far and you begin to challenge those limits politically, you may well have something to fear. Just look at the IRS monitoring of conservatives revealed recently.

What we need to be publicly debating, in bars, on talk radio, around dinner tables, in newspapers, is whether we want that future, because by my reckoning we have about five years to prevent it.

You might be a left-wing voter. You might hate conservatives. You might think it's a good thing that the IRS and other government agencies are spying on and harassing your political opponents. But think about something else for a moment: an infrastructure and political system that allows such outrages on one side, can equally be used against you when the time comes. Who, then, will you run to?

---

4 http://kellogg.nd.edu/faculty/fellows/dgroody/articles/TSSeptember09Groody.pdf

As German pastor Martin Niemoeller said:

First they came for the communists, and I did not speak out –
because I was not a communist;
Then they came for the socialists, and I did not speak out –
because I was not a socialist;
Then they came for the trade unionists, and I did not speak out –
because I was not a trade unionist;
Then they came for the Jews, and I did not speak out –
because I was not a Jew;
Then they came for me –
and there was no one left to speak out for me.

The whole "nothing to fear if you have nothing to hide" argument is presumptive. That is, it is made from a presumption that the State has the right to know your innermost thoughts or private dealings. It also presumes that you truly know the motives of the State, because if you actually don't know the motives, how can you possibly truly know if you have something to fear or not?

The whole point of "freedoms" is that they are seen as things the government has no right to touch. In the US Constitution, such freedoms are defined as given by God to all people, not given by the State. The reason for this was not because the founding fathers of the US were rampant Bible-bashers – far from it in many cases. What they were actually doing was creating a legal framework to say that "we the people" who are writing this constitution derive certain fundamental rights that the State should never be able to touch, therefore we allocate the giver of those rights as an authority higher than the State. If the State were to become the Supreme Being in a legal sense, it could cancel those fundamental rights at will. God is in the US constitution not for spiritual reasons but for legal reasons.

If you now approach issues from the presumption that the State reserves all rights to itself in the first instance, then you have already surrendered. Yet that is what is happening.

Take a look at 9/11. Under the guise of "national security", our daily lives have changed for the worse. Taking a flight is no longer a dream reminiscent of the TV series *Pan Am*, but a black comedy/farce.

How did it happen? A crisis presented itself and spin doctors leapt on the opportunity to "sell" the public a solution – tough new laws and regulations. Go back and look at the public debate at the time, we practically begged for it, begged for big government to make us safe.

The same sort of crisis talk and manufactured solution is rife in the climate change debate as well, but with the added twist that they've been so successful on the climate issue that a massive global governance framework is a mere whisker away from being implemented on the back of that Trojan Horse.

We are entering, I argue, what you could call the "post-democratic age", where the needs of the State are seen as trumping the needs of the citizens. The public exist, in the eyes of the elite, as a means of serving the State, not the other way around.

In a sense, this is not a new concept. It has been traditional throughout Western civilised history that the State was supreme, usually embodied in its monarch, and the people were subjects, not citizens.

The American and French revolutions in the 18th century put a spanner in that system of administration for a while, but supporters of the "supreme state" doctrine have regained the upper hand, even in the heart of America itself.

They have achieved it not through force but through propaganda and the persuasive powers of similar thinkers sprinkled as commentators and news editors in the mainstream media, they have done it with the assistance of idealists in the education systems who have ensured two generations of children have been indoctrinated into "thinking globally, acting locally".

We have become our own civilisational executioners, without even realising. We have delivered the bullets to be used against us, we have voted for those who will ultimately sell our freedoms down the river in return for a cushy retirement fund and big campaign donations.

The system is "post-democratic" in the sense that it is no longer true to the original principles. Sure, you still get the vote, but all elections seem to do is rearrange the deck chairs on the ship, while its unseen captains continue to steer their own course. Red, blue, left, right, nothing really changes, because the system looks after itself and throws up successful candidates who know the rules and know who they owe their success to.

Why is this? Who is really steering this ship? The answer, later in this book, may stun you.

Second century Roman satirist Juvenal coined the phrase (or at least he gets the credit for it), "give them bread and circuses". It was a cynical analysis of how the Roman powerbrokers manipulated the public with distractions and handouts and how the crowds loved it. Two thousand years later, have we really changed?

Intriguingly, Juvenal's *Satires* also gave us another phrase worth remembering: *quis custodiet ipsos custodies?* In English, "who watches the watchers?" An early forerunner of "power corrupts".

To understand where we've come from, and why powerbrokers want to take us back there, we will have to reacquaint ourselves with a little history. Above all, *Totalitaria* is a story about the drive to hold power over others. Do not assume for a moment that just because we drive flash cars and use smartphones, that the age old lusts and ambitions that have always driven human affairs are not still there, lurking just under the surface.

The essence of this book, then, is not conspiracy theory but "convergence theory" – the idea that sometimes a tipping point is reached because a range of different social threads begin to interweave amid a convergence of opportunities. The phenomenon is similar to a shark attack. The smell of the prey lures otherwise independent sharks in for a feeding frenzy. None are acting in concert, each has their own personal agenda, but they share a target: you.

As we shall discover, some groups in this story are motivated by economic power, some by political power, some by religious power and some by idealism. Some are motivated by a combination of these factors. To understand, we have to find out what drives them.

This book is split into two main sections. Firstly, an analysis of just how totalitarian-like our modern civilisation has become. What are the technological and social changes that encroach on our basic freedoms? What are the systems that might be acceptable to us today for one purpose, but which can be used against us by a future administration?

The second part of the book is devoted to finding out what drove our civilisation to this precipice, and meeting some of the people and groups behind it. Whereas part one is illuminating, I suspect you will find part two explosive as the book moves into top gear and accelerates to its conclusion.

Let's begin.

CHAPTER 1

The Facebook Generation

"You take the blue pill, the story ends, you wake up
in your bed and believe whatever you want to believe.
You take the red pill, you stay in Wonderland, and I
show you how deep the rabbit hole goes."
— Morpheus, The Matrix

Totalitaria to·tal·i·tar·i·a
noun \(|)tō-|ta-lə-|ter-ē-a\
: a state where the people are controlled in a very strict way with
complete power that cannot be opposed

There are, at the simplest level, three basic types of people in this world.

First, there are the 'conspiracy theorists'. These are the people who see design in all kinds of seemingly random events. "Join the dots," they warn, "and you'll see what's really going on". Some conspiracy theorists become so concerned at what they believe they've discovered that paranoia can set in, triggering that old joke: 'it's not paranoia if they *really* are out to get you'. Virtually everyone employed as a CIA analyst or FBI field agent is a conspiracy theorist.

Then there are the 'coincidence theorists'. These people can often be found writing newspaper columns or pontificating on a liberal talk radio station somewhere, playing down any nefariousness in business or politics. To the coincidence theorist, everyone has the best of intentions, everything in public life is done above board and for altruistic or sound commercial reasons, and we should all just let the government get on with it because, let's face it, we gave them a mandate by electing them. Coincidence theorists see design nowhere. Everything that happens is genuinely random. Things are what they seem, end of.

The third category of people are the ordinary punters going about their daily business. They're usually too busy living their lives to devote time to studying chicken entrails and old newspaper archives. If a conspiracy theorist makes a really good point, they'll give it some attention. They're just as likely to be swayed the other way by a persuasive, velvet-voiced coincidence theorist assuring them "there's nothing to see here, move on".

I suspect a lot of evolutionary psychological development has gone into fine tuning these categories of humans. After all, in the fight-or-flight world we used to live in, the paranoid were usually the first to run for safety. Sure, nine times out of ten they may have been spooked by nothing, but that tenth time was always a killer for those who stuck around.

Just look at birds of prey as you drive. The ones in the air, they're the conspiracy theorists. The pile of flattened feathers on the white line – that's a coincidence theorist. They missed the lesson on cause and effect.

The fact that conspiracy theorists usually survived to breed is probably a good indicator of why conspiracy theories are still as popular as they ever were. Most of us are prepared to give the occasional theory a slice of our attention.

Coincidence theorists, on the other hand, were traditionally the slower, more lumbering, antelopes in the herd. By the time they'd stopped laughing at the conspiracy theorists and finally realised there really was a lion in their midst, it was usually too late.

The truth, often, is found closer to the middle than either extreme.

Sometimes events just happen. The odds may be heavily against you winning the Lotto jackpot, but the odds also tell us that someone somewhere is usually in the right place at the right time.

Sometimes, however, events happen for a reason. They result from months, years or even decades of planning, and they happen because somebody wants a payoff of some kind. Usually, these somebodies don't wish to draw attention to themselves – particularly if they are jockeying for money or power. Like all humans, however, they are prone to bravado and speaking loosely to crowds they perceive as friendly.

The 20th century was the age when future historians will write that privacy died. Ironically, the sword that killed privacy is a double-edged one; the same technology opening up our lives is also helping expose some of those who'd prefer to stay hidden.

Some of the events in this book are the result of deliberate planning by two or more people. Some are just random coincidences that protagonists have taken advantage of. Some defy rational explanation altogether.

One of those random coincidences that big players have taken advantage of is Facebook.

In early 2011, Facebook welcomed its 600 millionth user, to great fanfare, ahead of its planned initial public offering. Sometime in 2013, Facebook welcomed its 1.15 billionth user, to no fanfare whatsoever.

Four hundred and twenty five million people rely on Google's email provider Gmail. Twitter has some half a billion users, four hundred million use Outlook.com and Dropbox has racked up some 200 million people whose mobile phone and computer documents are backed up to the Cloud storage system.

Every day there are literally billions of communications taking place over the internet, and there is an enormous amount to be said for the convenience of the information superhighway. The revolution is over, social media has won. Or has it?

With that level of exposure, you'd be hard pressed within your circle of friends not to find at least one of them (and probably a lot more) are on Facebook. The level of interconnectedness is legendary; everybody knows somebody.

Back in the early 1990s, a story began doing the rounds about "the Kevin Bacon phenomenon" – the strange fact that every Hollywood actor could somehow be linked to *Footloose* star Kevin Bacon within six degrees of separation, based on movies they'd appeared in. For example, Elvis Presley had appeared in a 1969 movie with Ed Asner, and Asner had two decades later appeared in a movie with Bacon, giving Asner a Bacon-separation of 1, and Elvis a Bacon-separation of 2, even though he and Bacon had never met.[5]

Sociologists have discovered the same thing in public life – that virtually all of us can be linked through our acquaintances and friendships,

---

5 Your humble author and Elvis Presley share a common *.grandfather several generations back, Andrew Presley of Paisley, Scotland in the early 1700s. Yet I have interviewed New Zealand singer John Rowles and Beatle George Harrison, both of whom met Elvis, giving me a "separation" of 2 as well.

within six degrees of separation, to anyone else on the planet – even to Kalahari bushmen.

This "interconnectedness" is one of the driving principles that Facebook is designed to exploit, but in doing so has it created a monster? Facebook is now officially cited as an aggravating factor in one in every five divorces. The opportunity to reconnect with old flames or new ones is more than many can resist.

While it can be argued infidelity is as old as the hills, it's certainly a lot easier now in this era of mass communication, than it was when people were confined to small villages for most of their lives as they were prior to the industrial revolution.

But just as it has made infidelity easier, Facebook has also made it easier to track, and that's where the Big Brother twist comes in. Just as you can detect and trace a partner or friend's online activities, so can others. To understand, though, how much your own privacy may be under threat from social networking, it's worth going back in time a little.

Former Israeli intelligence agent Ari ben-Menashe published a book called *Profits of War* in 1992 that caused a worldwide sensation, because it described for the first time a plan by Israeli spy agency Mossad to create a computer programme capable of monitoring individuals and working out who their circle of friends were. That programme was called PRO-MIS, and it was picked up not just by Mossad but also the CIA and the NSA in the US.

A commercial version of PROMIS was developed and sold to companies and government entities around the world, including New Zealand and Australia. Unbeknownst to clients, the software had a built in hack that allowed the CIA or other spy agencies to suck out the data on individuals that had been collected by, say, the New Zealand IRD or the local power or telephone utility.

"We can use this programme to stamp out terrorism by keeping track of everyone...All we had to do was 'bug' the programme when it was sold to our enemies," wrote ben-Menashe. "It would work like this: using a modem, the spy network would...tap into the computers of such services as the telephone company, the water board, other utility commissions, credit card companies etc. PROMIS would then search for specific information."

Information, said ben-Menashe, might include phone records that showed who the target associated with, or locations that he regularly visited. "The programme…would have the ability to track the movements of vast numbers of people around the world. Dissidents or citizens who needed to be kept under watch would be hard put to move freely again without Big Brother keeping an eye on their activities."

That was 1992. In those days, the "internet" was a dial-up entity reliant on telephone lines working at speeds of 2.4kbs, as against today's broadband speeds of 10 or 20 megabits per second or more. It was as slow as a wet week and very little substantive information could be exchanged. Nonetheless, PROMIS was state of the art. When the whistle was blown the intelligence agencies ran for cover, and organisations around the world that had purchased PROMIS ditched it. But the dream of tracking social networks via computer did not go away. Imagine the fun to be had if people voluntarily waived their rights to privacy and uploaded their personal information and friendships online for everyone to see.

"It'll never happen," muttered cynics in the spook community.

In 1998, the CIA ostensibly gave up the fight to lead the world in IT intelligence software.

"The leadership of the CIA made a critical and strategic decision in early 1998. The Agency's leadership recognized that the CIA did not, and could not, compete for IT innovation and talent with the same speed and agility that those in the commercial marketplace, whose businesses are driven by "Internet time" and profit, could. The CIA's mission was intelligence collection and analysis, not IT innovation," reports the CIA's own website.[6]

Instead of developing software, the CIA set up a company called In-Q-Tel in 1999 to go into venture partnerships with private capital firms to develop and fund promising new technologies that could help the CIA "data mine" for useful intelligence.

"In contrast to the remarkable transformations taking place in Silicon Valley and elsewhere, the Agency, like many large Cold War era private sector corporations, felt itself being left behind. It was not connected to the creative forces that underpin the digital economy and, of equal importance, many in Silicon Valley knew little about the Agency's IT needs. The opportunities and challenges posed by the information revolution to the Agency's core mission areas of clandestine collection and all-source analysis were growing daily. Moreover, the challenges are not merely from foreign countries but also transnational threats."

---

6 https://www.cia.gov/library/publications/additional-publications/in-q-tel/index.html

The CIA's website makes In-Q-Tel's mission clear:

"In-Q-Tel's mission is to foster the development of new and emerging information technologies and pursue research and development (R&D) that produce solutions to some of the most difficult IT problems facing the CIA. To accomplish this, the Corporation will network extensively with those in industry, the venture capital community, academia, and any others who are at the forefront of IT innovation."

The first CEO of In-Q-Tel was Gilman Louie, and he and his CIA subsidiary were appointed with seven others to the Board of Directors of the National Venture Capital Association of America in 2004. The chairman of the NVCA was James Breyer of Accel Partners, and one of his first tasks was to give a young student named Mark Zuckerberg US$13 million in venture capital.

What for? Well, in February 2004, Harvard student Mark Zuckerberg launched Facebook, a new social networking service designed to help fellow students stay in touch with each other.

James Breyer was also chairman of BBN, a Silicon Valley company whose Arpanet technology had helped kick off the internet with the assistance of the US Defence agencies. Joining Breyer at BBN in 2004 were both In-Q-Tel's Gilman Louie and Dr Anita Jones, a colleague of Gilman Louie's at In-Q-Tel, and a former advisor to DARPA, or the Defence Advanced Research Projects Agency.

In Facebook terms, these CIA and US Defence Department types were just one degree of separation from Facebook itself, and their chairman James Breyer was one of the initial bankrollers of Facebook. The first to see the potential was PayPal founder Peter Thiel, who kicked in half a million dollars of funding to Facebook just a couple of months after the project started. All of these people could see the potential stretched far beyond the "hot or not" photos of female students that Zuckerberg had pioneered.

One of DARPA's projects was "scalable social network analysis", or SSNA, which researcher Sean McGahan described this way: "The purpose of the SSNA algorithms program is to extend techniques of social network analysis to assist with distinguishing potential terrorist cells from legitimate groups of people ... In order to be successful SSNA will require information on the social interactions of the majority of people around the globe. Since the Defense Department cannot easily distinguish between peaceful citizens and terrorists, it will be necessary for them to gather data on innocent civilians as well as on potential terrorists."[7]

Think of the premise behind the hit TV series *Person of Interest*, and you will get the idea. Not so much of a system that can predict a fatality, but one that can predict potential fatality or "subversive activity" based on sophisticated algorithms. These programs – once they have enough access to background data from a population – can quickly identify and isolate disturbances and deviations from the average. Today, those algorithms are looking for terrorist cells. Two decades from now, who might be classified as a "terrorist" then?

As *New Zealand Herald* journalist Matt Greenop reported[8], DARPA and its new Information Awareness Office (whose official US government logo appears to have come straight from *The X-Files*) were planning to strip-mine the internet in search of whether ordinary citizens were somehow, through "friends of friends", linked to subversive or terrorist organisations.

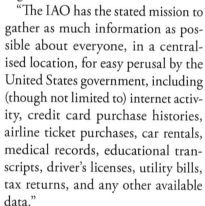

"The IAO has the stated mission to gather as much information as possible about everyone, in a centralised location, for easy perusal by the United States government, including (though not limited to) internet activity, credit card purchase histories, airline ticket purchases, car rentals, medical records, educational transcripts, driver's licenses, utility bills, tax returns, and any other available data."

If you've been paying attention, you will note the mission statement is remarkably similar to the former covert PROMIS project from the late eighties/early 90s. Only this time, US intelligence agencies have been able to do it largely in the open.

Well, not entirely in the open.

In July 2013, revelations from former CIA analyst Edward Snowden rocked the entire globe: intelligence agencies have been trawling virtually

---

7 Ethier, Jason. "Current Research in Social Network Theory". Northeastern University College of Computer and Information Science
8 "Facebook – the CIA conspiracy" by Matt Greenop, *NZ Herald*, 8 August 2007, http://www.nzherald.co.nz/technology/news/article.cfm?c_id=5&objectid=10456534

every email, every social media post and every phone call for traces of terrorist or subversive activity. They've been doing it through precisely the kind of backdoor programmes mentioned above – scalable social network analysis systems designed to harvest information out of the ether like a Dyson on speed.

Now, you may argue that if you have nothing to hide then what could possibly be wrong with data-mining of this nature. After all, we all know that messages over the internet can be intercepted and read. Well, yes they can, but most of us have approached that scenario over the past twenty five years on a law of averages basis: 'the chances of my communications being viewed by a spotty-faced 14 year old nerd in a dark bedroom in Chicago somewhere are slim to negligible, therefore I can be reasonably confident that of all the antelope in the herd at this moment on the internet, I'm not the one being singled out by lions'.

We call this the "it won't happen to me" scenario, and it's a behaviour we bring to all sectors of our lives. We know, for instance, there is a statistical risk involved in crossing the road. Every day around the world thousands of people die doing precisely that, but not one of us, anywhere on the planet, begins our day assuming it will be us.

The lesson from Snowden's document drop, however, is that it is us. We are all antelopes today.

Now, again, let's review this in line with the 'nothing to hide' meme. Today, the primary enemies of the state are Islamic terrorists and the like. Most of us will tick the box saying "I'm not one of those" and rightly assume that whatever they're harvesting from your emails and phone calls, "good luck to them". However, once again that doesn't take into account the longer term.

Firstly, is there a moral imperative for us to appear metaphorically naked in front of the State saying "search me" for the greater good? If you answer 'yes' to that question, how would you feel about a team of police turning up once a week to search your home and all the work you'd been doing that week, just to make sure you were not connected to terrorists? How would you feel about paying for all that surveillance through your taxes? Because you already are. The vast cost of trawling through all your internet data is being paid by you as the budgets for these massive programmes are increased each year.

In short, do you feel it should be normal or 'business as usual' for the State to assume that all citizens may be rogue and to treat them accordingly?

Secondly, even if you decide there is a moral imperative to treat everyone as a potential al Qa'ida threat in today's circumstances, do you agree that such powers should now be granted to the State for all time and all circumstances in the future?

What if, ten years down the track, a leader comes to power who you and many of your friends believe does not have your country's best interests at heart. Suppose that leader begins to treat political dissent as akin to treason, and because of his or her popular support (maybe they're a master of rhetorical and inspiring speech) they introduce laws restricting your rights to protest. Are you still of the belief that, in those circumstances, government agencies should have the power to trawl through your emails, phone calls and documents on your computer at will, with a view to finding out whether you meet their definition of 'subversive'?

Should such an administration have the power to open microphones and cameras in your home and effectively bug your conversations in the name of the State?

It doesn't matter whether such a leader is left-wing or right-wing. Once the infrastructure is in place, as it is today, and the public have by acquiescence given governments the power to use such infrastructure, it becomes very hard to roll it back short of a revolution.

This is the long term problem that arises from short term solutions. It is far too easy to come-up with a kneejerk solution to a pressing current problem, but bureaucracies by definition tend to grow like cancer and the bigger they get the more powerful they become in their own right. In the private sector, occasionally, we reach a point where companies become so dominant, like the Bell telephone company in the seventies, that we break them up into smaller units. No such facility exists to rein in Big Government.

Remember PROMIS from a few paragraphs back, with its "backdoor" software patch allowing intelligence agencies to raid users computers? Fast forward from 1992 to 2013 and that development is now on the record.

The NSA and Britain's GCHQ eavesdropping agency[9] have been using a system for the past ten years (roughly coinciding with the CIA's decision to set up a private sector IT company, In-Q-Tel) that "actively engages US and foreign IT industries to covertly influence and/or overtly leverage their commercial products' designs", according to leaked Snowden

---

9 Together with New Zealand's GCSB, Canadian and Australian intelligence

documents obtained by *The Guardian* and the *New York Times*.

The document doesn't name the technology companies whose products have been modified to aid the spy agencies, but it doesn't take a genius to work out the greatest intelligence interest is in those applications and websites with the highest traffic or usage.

One target, for example, is commercial file encryption. The NSA and GCHQ agents working in the private sector have figured out how to "insert vulnerabilities into commercial encryption systems".

The Guardian reports that these backdoors would be "known to the NSA, but to no one else, including ordinary customers, who are tellingly referred to in the document as 'adversaries'."

The Son of PROMIS is on the loose.

"These design changes make the systems in question exploitable through Sigint [signals intelligence] collection," says the NSA briefing paper, "with foreknowledge of the modification. To the consumer and other adversaries, however, the systems' security remains intact."

Among the systems already breached are the internet encryption standard "https" used to protect your financial transactions and password and login details, and voice-over-internet-protocol (VoIP) used for telephony and video-conferencing.

What does this mean? It means that government agencies can collect all your passwords and enter in to various websites posing as you. It means they can, by the same methods, go into your bank accounts online.

If a covert operation was set up to play with your sanity, item #1 on the list might be to clean your bank accounts out, or alternatively arrange for a sudden deposit from a bank in Medellin, Colombia, to appear in your account along with a polite covering note thanking you for assisting in the cocaine delivery!

The Snowden document sets out in clear terms the program's broad aims, including making commercial encryption software "more tractable" to NSA attacks by "shaping" the worldwide marketplace and continuing efforts to break into the encryption used by the next generation of 4G phones.

They probably won't have to work too hard, judging by this recent news report showing each mobile phone can be individually tracked:[10]

---

10 "Stanford researchers discover 'alarming' method for phone tracking, fingerprinting through sensor flaws", San Francisco Chronicle, 15 October 2013, http://blog.sfgate.com/techchron/2013/10/10/stanford-researchers-discover-alarming-method-for-phone-tracking-fingerprinting-through-sensor-flaws/

"One afternoon late last month, security researcher Hristo Bojinov placed his Galaxy Nexus phone face up on the table in a cramped Palo Alto conference room. Then he flipped it over and waited another beat.

"And that was it. In a matter of seconds, the device had given up its 'fingerprints'.

"Code running on the website in the device's mobile browser measured the tiniest defects in the device's accelerometer – the sensor that detects movement – producing a unique set of numbers that advertisers could exploit to identify and track most smartphones.

"It turns out every accelerometer is predictably imperfect, and slight differences in the readings can be used to produce a fingerprint (see below for a further explanation). Marketers could use the ID the same way they use cookies – the small files that download from websites to desktops – to identify a particular user, monitor their online actions and target ads accordingly.

"It's a novel approach that raises a new set of privacy concerns: Users couldn't delete the ID like browser cookies, couldn't mask it by adjusting app privacy preferences – and wouldn't even know their device had been tagged.

" 'I don't know if it's been thought of before', said Dan Auerbach, staff technologist at the Electronic Frontier Foundation. 'It's very alarming'."

In 2011, the power of social networking to shape geopolitics was laid bare in the Middle East uprisings. Both Facebook and Google proved crucial in mobilising networks of people instantly to force a revolution.

Coincidence? Who knows. But ultimately, as you enter your next Facebook update or disclose your newest friendship online, spare a thought for the supercomputers somewhere that are storing your details, who you associate with, where you live and what you like. If you are a good little Facebook user you will undoubtedly have entered photos of yourself so that identification and face recognition systems can pick you out of a crowd.

If you're under the impression that Facebook have improved their privacy settings after a series of controversies, well, yes, but only to a point. Facebook's terms and conditions make it clear that anything posted under the "Everyone" setting (which is the default setting when you first set up your account and details), is available to the whole world – possibly forever if the information is cached or taken by third parties.

If US intelligence agencies have figured out a way to tap into Facebook via the backdoor, as was planned with the original PROMIS software 19 years ago, then no amount of Facebook privacy settings is going to protect you from data mining. And the idea that spooks around the world are not tapping into the world's largest database of private social networking information, or trying to, is probably naive, given the mission statements of the US defence agency DARPA's Information Awareness Office and the CIA's In-Q-Tel.

Think about it for a moment. Virtually every website you login to these days has an option to login using your Facebook account. When you do that, Facebook stores the login and password information you have just entered. That information is added to Facebook's superprofile of yourself, potentially giving anyone with access to that information a built in dossier of what websites you have visited, and what your collection of passwords is.

George Orwell saw a time in his future where the State would force citizens to subject themselves to government monitoring through their TV sets, and history would be re-written by "the Ministry of Truth". He had no idea a time would really come when people would voluntarily give up what governments had been unable to force them to: the intimate details of their personal lives, social networks and much more, all for the buzz of seeing ourselves on a computer screen watched by 600 million others.

While the focus is confined to finding "terrorists" one could argue that those who have nothing to hide have nothing to fear. But ten years from now, with the global political landscape changing and the increasing likelihood of some kind of UN global governance structure to monitor individual travel and carbon use, what will a dossier of your political and social views, and the names of all your friends, be worth to someone, somewhere, if authorities want to shut down dissenting opinions fast?

The truth may be out there, but the question remains: whose truth?

CHAPTER 2

The Secret War For Control

*"Any system can be compromised given enough time.
We need an off switch, a back door, and this is our
last chance to build one."*
– Person of Interest

But it's not just US intelligence that might have an interest in knowing the whereabouts and friends-lists of Facebook users. In 2008 Chinese industrialist and People's Liberation Army frontman Li Ka-shing purchased a US$120 million stake in Mark Zuckerberg's Facebook.

Li is praised in business magazines as an entrepreneur. Indeed, a *Christchurch Press* editorial said just that when he tried to purchase the Port of Lyttelton in New Zealand a couple of years back:

"There is no need to be starry-eyed about the proposed venture. Li Ka-shing has risen from complete destitution as a refugee who fled the raping and pillaging of China by the Japanese in the 1930s to become a multi-billionaire.

"He did it by being an astute and hard-nosed businessman. He also did it, according to one account in a business journal, by 'remaining true to his internal moral compass' and operating with integrity."

But as *Investigate* magazine reported back then, things are never quite so Pollyanna simple:

"U.S. Congressman Dana Rohrabacher revealed that the U.S. Bureau of Export Affairs, the U.S. Embassy in Beijing and the Rand Corporation had identified Li Ka-shing and Hutchison Whampoa (Li's primary business) as financing or serving as a conduit for Communist China's military in order for them to acquire sensitive technologies and other equipment."[11]

Hutchison has since gone on to become a 50% owner of Vodafone Australia, a part owner of Vector Energy's Wellington power company in New Zealand and is investing in strategic assets throughout the West.

Pundits are predicting the 21st century to be Chinese-ascendant, and that means China and the US are locked in a covert battle at the moment for strategic supremacy. Most of that battle is being fought in the Pacific, where China has been massively boosting its diplomatic presence, donating huge sums in foreign aid to small Pacific Island states in return for business and military benefits.

The islands may seem small and insignificant, but as stepping stones across the Pacific they are just as strategic now as they were to the Japanese in World War 2. Collectively, the 14 Pacific Islands states cover a land area larger than Spain – hard as it may seem to believe when you look at them on a map. But the area of ocean they have economic and commercial rights to is a staggering 20 million square kilometres – larger than the total land area of the United States and Canada *combined*.[12]

"To most Pacific Island leaders, adopting a 'Look North' policy anchored on improved and closer relations with China was an inevitable progression," said a senior Fijian diplomat, Solo Mara. He told his audience the island states were tired of hearing "metropolitan powers" – a reference to Australia, New Zealand and the US – warning "for years" about China's "questionable security intent" in the region.

Describing China as a "sincere development partner", Solo Mara says the West lost a golden opportunity:

"China stepped in when other western development partners, such as the US and the UK, withdrew. Australia did not adequately fill the vacuum that was created – or one can say that they did not do it as effectively as the Chinese."

In 2007, Pacific Islands leaders travelled thousands of miles to visit the United States and were given what Mara called "a 10 minute photo opportunity" meeting with the then Secretary of State Condoleeza Rice. By 2012, the USA had picked up its game, sending Hillary Clinton to the Pacific Islands Forum meeting at Rarotonga. It was, argued Mara, "a clear

11 *Investigate* magazine citing Canada Free Press, http://www.thebriefingroom.com/archives/investigate_back_issues/april_06_issue/index.html
12 "Addressing the Future of Traditional and Non-Traditional Security in the Pacific Islands", speech by Solo Mara, Fiji High Commissioner to the United Kingdom, 6 March 2013, http://pacificislandssociety.com/speeches/fiji-high-commissioner-prepared-remarks-at-soas/

confirmation of Washington's realization that it must be more involved in the Pacific Islands or risk losing its influence entirely. It is interesting that Mrs Clinton was beaten to the islands by a multitude of senior Chinese Government officials – including Xi Jinping (now China's president)."

China has just launched its first aircraft carrier. It has undergone a bigger rearmament of its military than any other world power in recent history. In 2011 the Chinese navy opened fire on three Filipino fishing vessels off the coast of the Philippines. They were 100 miles from home, well inside the 200 mile territorial waters. More to the point, they were 600 nautical miles from the Chinese mainland. Regardless, the Chinese are claiming the whole South China Sea as Chinese territory.

A huge amount of the world's maritime trade goes through those waters, which link the Indian and Pacific oceans. Swedish MP Fredrik Malm says China's claim to own the entire sea would be like Russia telling all the countries bordering the Baltic Sea that they couldn't sail in it.

"Half of the world cargo traffic by sea, pass through [the South China Sea]. Chinese claims on this entire inland sea can result in far-reaching consequences not only for the free navigational rights of the sea, but also for the balance of power in the international system," explains Malm.[13]

"While the Arab world has witnessed uprisings and revolutions, and Europe and the United States is facing a severe debt crisis, a far reaching shift of power is occurring in East Asia.

"In the globalized world of today, continents, fates, and flows are increasingly intertwined. When fishermen from the Philippines are attacked by the Chinese navy it is a single incident in a long chain of events, which ultimately will affect Europe and Sweden. It will require an active and committed policy in order to prevent a future scenario where we stand irresolute to far more difficult situations."

You may be wondering what China's ability to choke world trade routes has to do with surveillance, but the connection is simple: there is a complex cold war being fought behind the scenes, and we are all unwitting participants. Part of the reason we have surveillance is to figure out when we are being played, before it's too late. Unfortunately, there are other more sinister reasons for surveillance too, and we will meet those in this book as well. This descent into totalitaria is by no means black and white; it's at least 150 shades of grey.

---

13 http://www.fredrikmalm2014.se/2011/12/chinas-alarming-rearmament.html

In 2012, a recent intelligence briefing for an international law enforcement agency fell into the lap of *Investigate* magazine, and it makes fascinating reading in the light of increasing Chinese activity in New Zealand, Australia and the Pacific, particularly with regard to immigration.

"Few of you will have heard about the "Sidewinder Report" begins the briefing.[14]

"Allowing it was tabled over a decade ago, after which money, influence and corruption were all brought to bear to have copies shredded, that isn't surprising. Fortunately a single digital copy survived, so we can still analyze/learn from this in-depth and rather alarming study, which is a very good example of Asian/Triad/Organized crime/long term planning.

"I personally believe a similar scenario exists/is being established in the likes of NZ and Australia, where similar immigration policies are in force. For this reason, I want to give you a detailed breakdown of the report, and you can perhaps reach your own conclusions.

"The report was commissioned in the mid 1990's codenamed "Sidewinder" and was a joint effort prepared by Canada's Secret Intelligence Service and the National Security Division of the Royal Canadian Mounted Police. Its mandate was to look at Chinese Triad involvement and integration into Canadian Financial and Governmental sectors.

"The report clearly found that over a period of time many Chinese triads, (Sun Yee On) agents of the Chinese Secret Intelligence Service (CSIS), and Hong Kong tycoons, had firmly established themselves in Canada and had acquired Canadian nationality.

"The two senior Canadian investigators were: Brian McAdam, a former diplomat who had uncovered the lucrative sale of Canadian visas during his posting at Canada's Hong Kong consulate. Canadian and Chinese consular staff were selling visas to members of the Chinese mafia and China's intelligence service, prices were as high as $100,000 per visa.

"The other was Michel Juneau, a former high-ranking French-Canadian intelligence officer who has spent over 20 years monitoring Chinese intelligence activities in the Asia-Pacific region. McAdam also noted that there was considerable political interference to shut down the investigation, and that it came from the highest levels. Former Prime Minister Jean Chretien's fall was indirectly attributed to his links to Chinese triads/corruption. He was also a key figure in CTIC, a Canadian Finance/

---

14 http://www.investigatemagazine.co.nz/Investigate/2655/china-trying-to-politically-infiltrate-nz-and-australia-report/

Loan institution which has since collapsed was in effect a pyramid type scheme with Triad backing.

"The saga continued under Chretien's successor. Canadian Prime Minister Paul Martin, was also part of the Chinese dynasty. He championed CIDA, The Canadian International Development Agency, which provides more development assistance to China than to any other country in the world. Once they get a hold, tentacles of Triad organisations go right to the top.

"The Canadian based triads operate on numerous fronts. Organised crime is also an area they are heavily involved in. Smuggling drugs into Canada for transport to the USA. They also use Canada as a key in the lucrative human smuggling or trafficking trade to Europe and the USA. They are directly aligned to COSCO, the Chinese international shipping company (which also operates in Australia/New Zealand) and over the years have obtained ownership of key companies throughout Canada.

"You will note locally many instances of takeover bids/company buyouts by Chinese consortiums; certainly not all are Triad affiliated ... or are they?

"Factors that were instrumental in the Triads gaining a foothold in Canada were: The Chinese outlook, plan long term for future generations – and Canadian Govt concession that allowed immigrants to settle in the country provided they put a large sum of money into Govt. bonds and brought or contributed to business. (Similar conditions apply in New Zealand)

"Although many Chinese then migrated to Canada, particularly Vancouver, obviously the question was not asked as to how they were able to raise the $500,000 required to gain automatic citizenship for them/their family. It only became obvious following 'Sidewinder' that much of that money was obviously Triad funded. The new Canadians were instructed to buy small companies preferably in downtown areas. Once a company was brought, the local name was not changed. That company was then used to buy additional land/businesses. To all intents and purposes, these sales were to native Canadians. IT companies, and those with government connections were key targets.

"They were instructed to make donations and get involved with political parties. Children studied hard and were directed at Government positions, many becoming well established in the ranks of the Immigration dept. [Name withheld] was Minister of [Portfolio withheld] during the [withheld]. He forged close links which China. "Somehow" he and his cronies are now all millionaires.

"By the year 2000, Chinese people affiliated to Triads owned one-third of downtown Vancouver. China invested over one billion dollars in 2001 to buy Canadian businesses in strategic areas and is also a large stockholder in Canada's Imperial Bank. It controls 15 corporations in the country's technology sector. By 2002, over 200 Canadian Companies were under the direct control of China's International Trust & Investment Corporation (CITIC).

"CITIC (Pacific) has many links to major Australian and NZ businesses. The Pengxin Group currently bidding to buy Crafar farms in New Zealand are linked to CITIC. CITIC operates directly under the general staff of the People's Liberation Army (PLA). It is also the world's largest private operator of container terminals, having lucrative stakes in 17 ports in Europe alone.

"Sidewinder found that significant amounts of arms, manufactured by a CITIC-controlled company, have been confiscated on Mohawk reserves. Vancouver is now considered the North American gateway for China's state-owned COSCO shipping company.

"Both U.S. Senate and Canadian intelligence sources have described COSCO as 'the merchant marine for China's military'.

"According to U.S. Intelligence reports, COSCO vessels do not just transport Oriental bric-a-brac. COSCO vessels have been caught carrying assault rifles into California and biological-chemical weapons components into North Korea, Pakistan, Iraq and Iran.

"Apart from this, Canadian law enforcement agencies have records of Chinese Triad criminal elements being active around all Canada's ports. Heavily involved through the whole 'Sidewinder' report is Li Ka-Shing ' known as Asia's most powerful man.

"Hong Kong Police asked Canada to investigate Li Ka-Shing back in 1988. Anne Marie Doyle, then Canadian High Commissioner denied that request. Li Ka-Shing is known to have strong Triad links," concludes the overseas intelligence briefing.

While the main example was the corruption of Canadian politics, similar things may already be happening in New Zealand. *Investigate* has broken stories of Chinese "businessmen" making big donations to the Labour and National parties, only to have the donations turn to scandal when alleged criminal links have been found.

Is it xenophobia to question the political and espionage motives of a very large superpower, or is "xenophobia" a racial smokescreen designed to distract from the real issue?

China has repeatedly engaged in economic and military espionage in the US, Canada, Australia and even New Zealand – where a breach of government computers was found to have originated in mainland China.

The Wikileaks cables show New Zealand diplomats are extremely concerned about China's activities, despite making reassuring noises to the New Zealand public.

NZ Ministry of Defence intelligence analysts told visiting US officials in 2006 that Chinese military infiltration of Pacific island states was posing "real security problems" for New Zealand, as the islands were "increasingly turning away from Australia and New Zealand to seek ties with Taiwan, China, Cuba and others."

One US briefing released by Wikileaks reveals New Zealand was worried about China fuelling the growth of "political instability in the Pacific Island nations."

Journalist Susan Merrell in the Pacific newspaper *Islands Business* reported how China had set aside a budget of US$375 million dollars with which to bribe Pacific Islands states to turn away from New Zealand and Australia. The newspaper reported that Chinese and Taiwanese bribery and competition had caused unrest and rioting in the Solomon Islands "after the election of Snyder Rini as Prime Minister. It was widely alleged that Chinese/Taiwanese bribes had secured him the position of PM."

New Zealand cables released by Wikileaks point similar fingers at the People's Liberation Army funding military aid to Fiji, Tonga and Papua New Guinea. "Equally troublesome are reported PLA links to paramilitary forces in Vanuatu," noted a 2006 cable.

China's tricky behind the scenes game with New Zealand included hiding the fact that one of its aircraft was using New Zealand as a stopover point on a secret mission to fly senior Chinese leaders to meet Fiji's Colonel Bainimarama in Suva. A Wikileaks cable from US diplomat Dan Piccuta states: "The Chinese sought to obscure plans for [Vice President] Xi's stop in Fiji by omitting the onward destination of Xi's aircraft in the Chinese Government's application to the New Zealand Government to transit New Zealand airspace."

Given that New Zealand has to rely on official Chinese assurances about the character of business people and migrants seeking to live in New Zealand, the relationship between the two countries requires a lot of trust. Clearly, behind the scenes, there's very little trust. Which leaves the question, just who has been allowed to slip into New Zealand as part of long-term strategic moves by China over the past 25 years, and what's really behind the Chinese investment juggernaut?

The questionmarks hang over China's dealings with the other western nations. The Operation Sidewinder report shows Canada's security has been heavily compromised, and America too is being squeezed. The Chinese have turned economic espionage and sabotage into an art form. Take America's foreign debt mountain. The Chinese own US$1.3 trillion in US debt, while the Japanese hold $1.1 trillion. And that's just their holdings in US Treasury bonds. That does not include total investment in companies or private financial holdings.[15]

A Pentagon report in 2012 suggests the debt is more of a problem to China than it is for the US, and to some extent it could be America's way of keeping the sleeping dragon on a short leash. China's economy depends on the US, for now, so anything that harms the US will also harm China. At least, that's the supposition.[16] Tell that to the cyber security warriors tracking Chinese cyber attacks, however.

Chinese telecommunication giant Huawei makes mobile phones and network infrastructure. Under New Zealand's free trade agreement with China, Huawei has ended up as a contracted provider for New Zealand's high speed broadband infrastructure, and New Zealand can't shut Huawei out without jeopardising the entire free trade deal.

However, in cold black and white terms there are some who see Huawei and other Chinese giants as a serious security threat.

In May 2013, the *Washington Post* broke a story that Chinese government hackers had managed to steal the blueprints for Australia's new intelligence agency headquarters. It wouldn't be the first embarrassing blueprint gaffe in espionage history – the brand new US embassy in Moscow famously had to be torn down in the 1980s after the CIA finally realised that the Soviet labourers who'd been working on the site were KGB operatives. Apparently the possibility had not occurred to the Americans. When they checked the found the entire embassy was so "riddled" with bugs it was like a case of electronic termites.[17]

It's not as though the Soviets didn't have prior form. In 1946 the Rus-

---

15 http://www.washingtonpost.com/blogs/worldviews/wp/2013/10/10/this-surprising-chart-shows-which-countries-own-the-most-u-s-debt/
16 http://www.forbes.com/sites/kenrapoza/2013/01/23/is-chinas-ownership-of-u-s-debt-a-national-security-threat/
17 Henry Hyde, Congressional Record, 26 October 1990, http://www.fas.org/irp/congress/1990_cr/h901026-embassy.htm

sians presented the US ambassador to Moscow with a copy of the US "Great Seal" to hang in the American embassy. For seven years until it was discovered, the Ruskies were able to eavesdrop on embassy conversations using a special bugging device they had planted in the Seal. Of course, the dastardly plan all hinged on where the Ambassador chose to hang the Seal, but the Russians probably figured it was a safe bet the prestigious icon wasn't going to be displayed by the men's urinal.

China may be in the news for all the wrong reasons now, but they too have been on the receiving end of some western espionage stunts. In 1990 they discovered the Chinese Embassy in Canberra, Australia was being bugged by Australian and US intelligence[18], and in 2001 China took delivery of a brand new Air Force One for President Jiang Zemin – a Boeing 767 – only to find no fewer than 27 electronic eavesdropping devices hidden inside the plane.

"The aircraft has been sitting on a military airstrip north of Beijing, unused with much of its upholstery and many of its fittings ripped out, since October when Chinese test pilots detected a strange and unfamiliar whine emanating from its body," reported the *Telegraph* in London.[19]

"A search of the twin-engined aircraft, which was manufactured and fitted out in America, yielded 27 devices, according to Chinese officials, hidden in its seats, lavatory and panelling."

Precisely what conversations the Americans were hoping to hear in an aircraft toilet is unclear, but one bug was also found in the headboard of the president's bed on the plane.

Diplomacy seems to be the only game in town where it is compulsory to destroy the gifts you've been given, whilst maintaining a forced smile for the outside world at all times. Even Britain's Queen Elizabeth wasn't immune, nor amused.

"An electric teapot presented to the Queen by the Russians has been removed from Balmoral as a security precaution," reported the *Telegraph*.[20]

It wasn't exactly the most inconspicuous of teapots. This one was 60cm (two feet) long and coloured a garish red and yellow, but MI5 felt its wiring

18   http://web.archive.org/web/20090503131918/http://www.diggerhistory.info/pages-conflicts-periods/ww2/pages-2aif-cmf/aib-asio.htm

19   "China finds spy bugs in Jiang's Boeing jet", *The Telegraph*, 20 January 2002, http://www.telegraph.co.uk/news/worldnews/northamerica/usa/1382116/China-finds-spy-bugs-in-Jiangs-Boeing-jet.html

20   "Russia's teapot gift to Queen 'could have been bugged'", The Telegraph, 25 November 2008, http://www.telegraph.co.uk/news/uknews/theroyalfamily/3515927/Russias-teapot-gift-to-Queen-could-have-been-bugged.html

may contain some kind of listening device. Unfortunately, the *Get Smart* device, known as a samovar, had been sitting in the Queen's drawing room for nigh on 20 years since she'd received it from a Russian sports team, so British intelligence was not really on the ball.

Of course, it may have been simply a teapot. That's the point about espionage – you can never really tell. For all we know the Americans planted 27 obvious bugs on Jiang Zemin's jet purely for the amusement of watching the Chinese rip their new plane to shreds, whilst hiding some undetectable devices elsewhere in the plane.

In response to this year's news that Chinese hackers had stolen the plans for Australia's spy headquarters, the question was once again raised about the wisdom of allowing Chinese telecommunications manufacturers to help build your communication infrastructure.

Huawei's reaction, through its head of security John Suffolk, told the media to get real – Governments everywhere spy on everyone:[21]

"Governments have always done that," Suffolk said. "The harsh reality is every government around the world has a similar strap-line for their security agencies…Some people say that spying is the second-oldest profession, where people have tried to get information off us for somebody else, so I don't think anyone is surprised that any government around the world is trying to find out what other governments around the world are doing.

"Governments have to really focus on what quiet steps they're going to take, accepting no government will really trust 100 per cent another government, regardless of the laws, the policies and procedures," he added.

A bugging device in the Queen's teapot, however, is a million light years away from a situation where every phone call you make, every email you send and every webpage you visit is going through cables and servers and comms equipment built by a country you perceive as a threat.

Huawei was established in 1987 by a former Chinese Red Army soldier, at a time when the country was staunchly communist. While it remains a one party state, China has liberalised its economy and travel restrictions. Huawei has gone on to become the world's largest supplier of telecommunications equipment.[22]

---

21 "Chinese Firm Huawei: Sorry, But We're All Going To Have To Get Used To Government's Spying On Us", Agence France Presse, 29 May 2013, http://www.businessinsider.com.au/chinese-firm-huawei-sorry-but-were-all-going-to-have-to-get-used-to-governments-spying-on-us-2013-5
22 "At Huawei, Matt Bross Tries to Ease U.S. Security Fears", by Ashlee Vance and Bruce Einhorn, Bloomberg Business Week, 15 September 2011, http://www.businessweek.com/magazine/at-huawei-matt-bross-tries-to-ease-us-security-fears-09152011.html

The company employs 140,000 people worldwide and supplies 45 of the world's top 50 telecom providers, including all the major networks in New Zealand, but suspicions have grown that – like the Li Ka Shing operations – things are not all as they seem. For its part, Huawei has hit back, inviting British security to go over their software and hardware with a fine tooth comb. Given MI5's failure to detect a teapot for 20 years, no one is sure whether the failure to find anything suspicious in the Huawei gear is a good sign or not. Additionally, just because the software is clean 'today' doesn't mean it will be after one of its automatic 'upgrades' down the track.

For all of its talk of openness, Huawei had never disclosed its board of directors until 2010, and hasn't published details of its shareholders. It denies being a front for the Chinese Army, but it's hard to know.

A proposed new fibre-optic cable linking the United States, New Zealand and Australia was put on ice after American objections to Huawei's involvement in the project.

One of the US concerns is that Huawei equipment may contain a 'back door' allowing Chinese intelligence to access it once it's installed in Western countries. Given what we now know about Western intelligence doing precisely that themselves, it seems to be a reasonable, if somewhat hypocritical, assumption. As Huawei's John Suffolk remarked, get over it, everyone is doing it.

A US Defence Department briefing in 2008 listed Huawei as a company with "close ties to the PLA [People's Liberation Army]" and that it cooperates closely with the Chinese military on research and development. This has been a policy issue for China, with the then president Hu Jintao highlighting the rapid technology developments available to the PLA from partnerships with commercial IT companies in China.[23]

Because of China's one party structure, however, cooperation may not so much be voluntary as mandatory. The Asia Society in one report says people – in this case hackers – get leaned on to share what they discover with the authorities:

"The Chinese government has employed this same tactic in numerous intrusions. Because their internal police and military have such a respected or feared voice among the hacking community, they can make use of the

---

23 "Military Power of the People's Republic of China" US Defense Dept, 2008, http://www.defense.gov/pubs/pdfs/China_Military_Report_08.pdf

hackers' research with their knowledge and still keep the hackers tight-lipped about it. The hackers know that if they step out of line they will find themselves quickly in a very unpleasant prison in western China, turning large rocks into smaller rocks."[24]

If you think it's all paranoia, consider the case of the humble Insignia Digital Photo Frames. These imports from China featured in vast numbers of homes in 2008, but were later discovered to have a bug in their software that allowed them to raid a householder's computer and upload data back to a server in China. The bug was not detectable by commercial anti-virus programmes. US retail giant Best Buy yanked the offending digital photo frames off the shelves.[25]

An audit of US military equipment found sophisticated Chinese counterfeiting of parts for military helicopters and maritime surveillance aircraft, and that these "parts" had been installed on a number of US military units. There's no word on what those parts did, but the fact that they ended up on intelligence-gathering aircraft is a sign of how hi tech the cold war now is.[26]

Of course, all this spills over into your own daily life.

Debate around the security of webcams and microphones in computers, tablets and smartphones continues to rage, but from the "what-do-they-know-that-we-don't?" department, it is probably relevant that the NSA is advising its employees to disable their webcams on their Apple devices:[27]

"The best way to disable an integrated iSight camera is to have an Apple-certified technician remove it. Placing opaque tape over the camera is less secure but still helpful. A less persistent but still helpful method is to remove /System/Library/Quicktime/QuicktimeUSBVDCDigitizer. component, which will prevent some programs from accessing the camera.

"To mute the internal microphone, open the Sound preference pane, select the Input tab, and set the microphone input volume level to zero. To disable the microphone, even if it means crippling the sound system, remove the following file…"

Usually switching on someone's webcam firstly requires Remote Admin-

24 "The Chinese Cyber Threat" by Brian Le, Asia Society, 4 August 2011, http://asiasociety.org/blog/asia/chinese-cyber-threat
25 "Obama Invokes Cold-War Security Powers to Unmask Chinese Telecom Spyware", Bloomberg, 1 December 2011, http://www.bloomberg.com/news/2011-11-30/obama-invokes-cold-war-security-powers-to-unmask-chinese-telecom-spyware.html
26 ibid
27 http://www.nsa.gov/ia/_files/factsheets/macosx_hardening_tips.pdf

istration Tool (RAT) software to be loaded onto the target's device. For the general public, that can involve Trojans buried inside downloaded programmes, pictures and webpages, or it can involve a wireless network breach or some other more direct attack. If a state agency is doing the hacking, more cloak and dagger techniques may be used to get the RAT loaded on your system. Once there, however, the outcome is the same.

The irony of all this is that, by the time you reach the end of this book and its revelations about the totalitarian agenda, the Chinese may well look like the good guys.

CHAPTER THREE

Control Of Your Home

"I know you're out there. I can feel you now. I know that you're afraid. You're afraid of us. You're afraid of change. I don't know the future. I didn't come here to tell you how this is going to end. I came here to tell you how it's going to begin. I'm going to hang up this phone, and then I'm going to show these people what you don't want them to see. I'm going to show them a world without you. A world without rules and controls, without borders or boundaries; a world where anything is possible."
– Neo speaks to the Machine, The Matrix

Perhaps one of the biggest manifestations of the battle for control of the world is the fraught and hotly-debated issue of climate change. According to this doctrine, planet Earth is being irrevocably overheated by the activities of humans, and we are now facing a period of runaway global warming in which millions of small furry animals will perish and sea levels will rise – Noah-like – until we all drown.

The only way to avoid such a fate, say the totalitarians, is to regulation-bomb civilisation back to the dark ages, to a point where we have a zero carbon footprint. Regulations would include "carbon accounts" for every human where, if you exceeded your monthly allocation, your car or house could be remotely shut down by the State unless you paid further fees.

One of the technologies introduced to enable this kind of micro-control is "smart meters" for electricity and gas supply to the home. Like all good totalitarian initiatives, officials find the public much easier to herd towards the desired outcome if they pitch the initiative right. In this case, the pitch is that smart meters will save you money and make your homes more automated.

The principle they work on is simple. Instead of the old-fashioned analogue meter boxes, a smart meter is computerised and can communicate with modern appliances in your home. These can include computer networks, refrigerators, dryers, washing machines, security systems, TVs, heating systems and the like. Instead of being read by a meter company once a month, smart meters use a built-in mobile phone to send data on your electricity usage back to the power company every day – in fact, up to four times every hour.

Built in to this power company computer on your outside wall is the ability for two-way traffic. Not just does your unit report on everything you are using, but it can also accept instructions from head office. If the power company chooses to switch off one of your appliances, or even your whole house, it can do so remotely thanks to the sim card in the smart meter.

As many critics have now pointed out, such a system is a glaring security weakness. If the State can shut you down via remote computer codes phoned through to your home, so too can hackers. They may be random, autonomous nerds without girlfriends – seeking to impress their nerd mates – or they could be based in Beijing's hacker quadrant – the city block where agents employed by the Chinese government breach the security systems of the US, UK, Australia, New Zealand and Canada with monotonous semi-regularity.

A one-off power cut caused by such activity might be inconvenient, but a power cut instituted nationwide during a conflict or even a mere trade dispute, as a result of a built-in Trojan horse flaw, is entirely another.

Equally, even if life doesn't get that dramatic, the smart meters could allow criminals with friends in high places to shut down your home alarm system to enable your premises to be burgled. The possibilities are endless.

If you think this is just journalistic scaremongering, think again. Industry analysts like British cyber security specialist Professor Ross Anderson are all over this one.[28]

---

28 "Who controls the off switch?" Anderson R & Fuloria S, http://www.cl.cam.ac.uk/~rja14/Papers/meters-offswitch.pdf

"The off switch creates information security problems of a kind, and on a scale, that the energy companies have not had to face before. From the viewpoint of a cyber attacker – whether a hostile government agency, a terrorist organisation or even a militant environmental group – the ideal attack on a target country is to interrupt its citizens' electricity supply. This is the cyber equivalent of a nuclear strike; when electricity stops, then pretty soon everything else does too. Until now, the only plausible ways to do that involved attacks on critical generation, transmission and distribution assets, which are increasingly well defended. Smart meters change the game."

You would think western governments would have taken this into consideration, but if they have there is no real sign of it. Europe, the USA, New Zealand and Australia are diving headlong into building so-called "smart grids". In the US, it was put on the agenda with the Energy Independence and Security Act of 2007, then catapulted into the spotlight when President Obama signed the American Recovery and Reinvestment Act of 2009 and pumped US$4.5 billion into developing the smart grid.[29] Some estimates put the cost now at over $11 billion.

Reports have also called smart meters the "key to national efforts to further energy independence and curb greenhouse gas emissions"[30].

In Europe, the target for a smart grid switchover is 2022. Some countries, like the Netherlands, jumped on the bandwagon early, proposing to make smart meters mandatory by 2013. Under the Dutch legislation, refusal to abide by this state directive was to be punishable by a 17,000 euro fine or jail for up to six months. The "compulsory" aspect was ditched after a widespread public revolt in Holland. It could yet re-emerge as an issue, because the European Union has directed a "mandatory" switch of all households to smart meters once a cost-benefit analysis has proven positive in Europe's eyes. European officials are being supported in their compulsion approach by academics who say smart meters are only economic for a country if every householder is required by law to install them:

"The establishment of a valid legal obligation, either at the European or national level, might serve as the clearest, safest, and most sustainable way of securing successful implementation."[31]

---

29 M LaMonica, "Obama signs stimulus plan, touts clean energy", CNN,
Feb 7 2009, at http://news.cnet.com/8301-11128_3-10165605-54.html
30 NIST, Framework and Roadmap for Smart Grid Interoperability Standards Release 1.0 (Draft), Sep 2009
31 Rainer Knyrim and Gerald Trieb, "Smart metering under EU Data Protection Law", *International*

Bearing in mind the totalitarian intention to wheel these things in come hell or high water, Professor Anderson's security analysis concludes:

"Electricity and gas supplies might be disrupted on a massive scale by failures of smart meters, whether as a result of cyberattack or simply from software errors. The introduction of hundreds of millions of these meters in North America and Europe over the next ten years, each containing a remotely commanded off switch, remote software upgrade and complex functionality, creates a shocking vulnerability. An attacker who takes over the control facility or who takes over the meters directly could create widespread blackouts; a software bug could do the same.

"Regulators…have started to recognise this problem. There are no agreed solutions as yet; in this paper we've discussed the options. Possible strategies include shared control, as used in nuclear command and control; backup keys as used in Microsoft Windows; ratelimiting mechanisms to bound the scale of an attack; and local-override features to mitigate its effects. It's important that these issues are discussed now, before large-scale rollout creates large-scale vulnerabilities in too many utility areas (and indeed national markets)."

The smart meter industry meanwhile has come out swinging, dismissing privacy fears around smart meters as "paranoia":

"Americans are already aware that life in a digital world involves trade-offs between convenience and privacy," says Patty Durand, the executive director of America's Smart Grid Consumer Collaborative.[32]

"That calls for discussion and debate of those trade-offs and might need to include whether the trade-offs apply to smart metering data. Living in a digital world and understanding and balancing these trade-offs would seem to require more robust engagement among the relevant parties, rather than indulge in fear and a search for bogeymen.

"While the constitutionality and legality of the NSA surveillance needs to be openly reviewed, let's take a deep breath and consider whether this is a game changer for power utilities and their efforts to run the grid more efficiently and with a greater degree of environmental responsibility.

"Placed into context, the use of smart meters to enable more sophisticated energy use by the consumer and more efficient, sustainable grid-related practices by utilities makes at least as much sense as other tech-

Data Privacy Law, March 1, 2011, p 128
32 "The 'surveillance state' and smart meters" by Patty Durand, Intelligent Utility, 18 July 2013 http://www.intelligentutility.com/article/13/07/surveillance-state-and-smart-meters

nologies where the trade-offs between value and privacy are less clear."

In other words, says Durand, the ends ('sustainable' energy) justify the means (a potential risk to consumers). Not that she sees a real risk.

"Utilities and public utility commissions remain responsible for robust data privacy practices and policies with sufficient transparency to allay their customers' concerns. The movement toward offering opt-out policies, with corresponding fees, is one step in that direction. An iron-clad policy that any personally identifiable information derived from smart metering is the sole property of the account holder is another step, which is already widely embraced.

"We're really at the beginning of important conversations about how to live in a digital world while retaining a sense of privacy. Let's recognize that's where we are. The power industry and its customers have yet to grasp, fully explore and, perhaps, resolve related issues involving smart meters.

"News of the NSA's domestic surveillance of electronic communications should drive that engagement rather than plunge us into a murky world of pervasive paranoia. Many people have worked hard to assure consumers that their energy data is safe as always."

Durand may be walking the talk, but is her confidence in "robust" data privacy practices by power companies misplaced? The evidence suggests so:

"Cyberspies have penetrated the U.S. electrical grid and left behind software programs that could be used to disrupt the system, according to current and former national-security officials," the *Wall Street Journal* has reported.[33]

"The spies came from China, Russia and other countries, these officials said, and were believed to be on a mission to navigate the U.S. electrical system and its controls. The intruders haven't sought to damage the power grid or other key infrastructure, but officials warned they could try during a crisis or war.

" 'The Chinese have attempted to map our infrastructure, such as the electrical grid', said a senior intelligence official. 'So have the Russians'.

"The espionage appeared pervasive across the U.S. and doesn't target a particular company or region, said a former Department of Homeland Security official. 'There are intrusions, and they are growing', the former official said, referring to electrical systems. 'There were a lot last year'.

---

33 "Electricity Grid in U.S. Penetrated By Spies", by Siobhan Gorman, WSJ, 8 April 2009, http://online.wsj.com/article/SB123914805204099085.html

"Many of the intrusions were not detected by the companies in charge of the infrastructure."

Yup, that last line instils confidence alright.

In February 2013, another power utility hit the headlines when the private data of 110,000 customers was accessed:[34]

"Central Hudson Gas and Electric based in Poughkeepsie, New York is working with state and federal authorities and industry groups to investigate a cyber attack earlier this month where hackers gained entry to as many as 110,000 customer accounts. Employees detected the computer system intrusion, which happened over a weekend, as a result of regular control procedures, the utility said.

"So far there appears to be no evidence that customer information was misused or downloaded during the incident, but the utility has warned customers to monitor bank accounts for suspicious or unauthorized activity."

Department of Homeland Security has also warned that smart meters and grids provide a range of opportunities, not just a national security issue but a household security one as well, where networks can be "accessed with minimal skills in order to trespass, carry out nefarious activities, or conduct reconnaissance activities to be used in future operations."[35]

A 2011 hacking conference in the US learned that a $60 piece of software available off the shelf was capable of bringing smart grids to their knees.[36] Then there's the flipside – the discovery that smart meters are so easy to hack that even homeowners are getting in on the act:

"The FBI has seen an increase of smart meter hacks which allow consumers to reduce power bills by 50-75%. Crazy hacking skills are not required and can be accomplished by using a magnet to fake readings or hiring hackers to attack smart meters. The FBI warned the cost of smart meter fraud may cost utility companies $400 million per year."[37]

The sheer scale of the data about private households that can be assem-

34 "Yes it can happen here: Poughkeepsie utility hacked", Smart Grid News, 27 February 2013, http://www.smartgridnews.com/artman/publish/Technologies_Security/Yes-it-can-happen-here-Poughkeepsie-utility-hacked-5553.html#.UhlMbJJHLXM
35 DHS security bulletin, 16 September 2011, uploaded to internet at http://publicintelligence.net/ufouo-dhs-bulletin-anonymous-hacktivist-threat-to-industrial-control-systems-ics/
36 "Power grid cybersecurity: $60 piece of software could bring mass chaos" by Darlene Storm, Computerworld, 14 November 2011, http://blogs.computerworld.com/19270/power_grid_cybersecurity_60_piece_of_software_could_bring_mass_chaos
37 "FBI Warns Smart Meter Hacking May Cost Utility Companies $400 Million A Year", Network World, 10 April 2012, http://www.networkworld.com/community/blog/fbi-warns-smart-meter-hacking-may-cost-utilities-400-million-year

bled from smart meter analysis is contained in a report for the EU:[38]

"Smart meter data, when measured in intervals of 4 hours, exactly reveal when a person is at home, when he is sleeping and when he is preparing his meals. When using shorter intervals, of minutes or seconds, electric devices can be identified on the basis of use profiles, such as a fridge, coffee machine, washing machine, toaster, microwave, and TV. These data can reveal if someone eats a cold or a hot breakfast, when laundry is done, or whether the kids are alone at home. It is even possible to determine which channel a TV is tuned to, through an analysis of the broadcast programs, particularly if the TV is tuned to a longer program such as a movie. The interfering noise in the meter data of other energy-consuming devices can most likely be filtered out in case movies are watched of 90 minutes or longer.

"This demonstrates that the more detailed smart meter readings are, the more privacy-sensitive the data become. Real-time readings in intervals of minutes can reveal many details of home life and paint a disturbingly clear picture of people's behaviour and preferences."

In early 2012 the then-director of the CIA, David Petraeus, boasted of his agency's ability to spy on persons of interest through their smart meters and home appliances, or what is known in tech circles as "the Internet of Things":[39]

"'Transformational' is an overused word, but I do believe it properly applies to these technologies," Petraeus told investors in the CIA's venture capital company In-Q-Tel,[40] "particularly to their effect on clandestine tradecraft."

"Items of interest will be located, identified, monitored, and remotely controlled through technologies such as radio-frequency identification, sensor networks, tiny embedded servers, and energy harvesters – all connected to the next-generation internet using abundant, low-cost, and high-power computing," Petraeus said.

The only redeeming irony for privacy advocates was Petraeus' forced resignation as CIA director a few months later after being caught in a

---

38 C. Cuijpers & B.J. Koops (2012), 'Smart metering and Privacy in Europe: Lessons from the Dutch Case', in: S. Gutwirth et al. (eds), European Data Protection: Coming of Age, Dordrecht: Springer, p. 269-293

39 "CIA Chief: We'll Spy on You Through Your Dishwasher" by Spencer Ackerman, Wired.com, 15 March 2012, http://www.wired.com/dangerroom/2012/03/petraeus-tv-remote/

40 As noted elsewhere in this book, In-Q-Tel was the CIA's backdoor into Facebook, Google and other IT giants for the purposes of surveillance

clandestine affair. Even those pushing the 'nothing to hide' doctrine appear to have reason to fear.

A German security conference heard how hackers had obtained detailed data on a householder's habits thanks to a smart meter:[41]

"They showed that the type of LCD TV set could be identified, what TV program was on, or if a movie was playing from a DVD or other source. The research team called for a tightening of data protection regulations. Building upon that, the 28C3 presentation "Smart Hacking For Privacy" demonstrated that consumers can be identified via the data collected by a smart meter, from the types and amount of your devices, your TV shows, to scanning for copyright-protected (or pirated) movies being watched."

The question for consumers, then, is a simple one: when the Government says 'trust us, we know what we are doing', do you believe them? If the answer is 'yes', then ask whether you trust their motives.

The other problem is trying to second guess what sort of futuristic doors existing Trojan horses will open up down the track. This is "the wedge" principle. Does equipping homes with smart meters, for example, mean that sometime down the track government agencies or hackers will be able to remotely open the webcam on your flash new 42 inch interactive TV in your bedroom?

It probably won't come as surprise to some that things like this are already happening, just using good old-fashioned computer hacking and Trojan software.

"Webcam hacking has officially gone mainstream with yesterday's revelation that the new Miss Teen USA, Cassidy Wolf, was the victim of a 'sextortion' plot in which someone slipped Remote Administration Tool (RAT) software onto her computer and used it to snap (apparently nude) pictures of Wolf in her room. 'I wasn't aware that somebody was watching me (on my webcam)', she told *The Today Show*. 'The light (on the camera) didn't even go on, so I had no idea'."[42] The first she knew of the hack was when a half-naked picture of her suddenly became the avatar for her Twitter account – definitely a giveaway of skulduggery – and she then found her account passwords had been changed.

In the words of Mulder and Scully in the *X-Files*, "trust no one".

41 http://www.networkworld.com/community/blog/hacking-privacy-2-days-amateur-hacker-hack-smart-meter-fake-readings

42 "Webcam spying goes mainstream as Miss Teen USA describes hack", Arstechnica, 17 August 2013, http://arstechnica.com/tech-policy/2013/08/webcam-spying-goes-mainstream-as-miss-teen-usa-describes-hack/

You might be asking the logical question, 'why would the Government want to spy on us in our beds?', and of course the correct answer probably is that Homeland Security don't actually want to do that, but a bored 22 year old who can piggy-back into your home on the back of that kind of infrastructure might.

On the other hand, the ability to switch on your laptop camera, or your phone microphone anywhere in your home, might be of use to government agencies in the future should they suspect that you, like air traveller Shoshana Hebshi of 'spread 'em and cough' fame, have been seen with the wrong kind of people.

The question you should ask yourself, if you have nothing to hide, is whether you prefer to live in a glass castle with no curtains, or whether you regard your home as a sanctuary from the outside world?

On precisely that note, how many people are aware that home appliances are already being designed that will literally keep an eye on you for big media companies?

"National telecommunications companies are exploring technology for digital video recorders (DVRs) that would record the personal activities of consumers as they watch television from the privacy of their own homes," warn US congressmen Mike Capuano (D-MA) and Walter Jones (R-NC) in a joint news release this year. The two men from different sides of the political divide are introducing a bill trying to ensure minimum privacy safeguards are adhered to before the technology goes on sale.[43]

"Current law is silent on these devices and this legislation would require both an opt-in for consumers and an on-screen warning whenever the device is recording information about consumers.

"This may sound preposterous but it is neither a joke nor an exaggeration. These DVRs would essentially observe consumers as they watch television as a way to super-target ads. It is an incredible invasion of privacy. Given what we have recently learned about the access that the government has to the phone numbers we call, the emails we send and the websites we visit, it is important for consumers to decide for themselves whether they want this technology. Think about what you do in the privacy of your own home and then think about how you would feel sharing that information with your cable company, their advertisers and your government," states Capuano.

---

43 Press release by Congressmen Capuano and Jones, June 2013, http://www.house.gov/capuano/news/2013/pr061313.shtml

"Allowing this type of technology to be installed in the homes of individuals without their consent would be an egregious invasion of privacy," adds Jones. "When the government has an unfortunate history of secretly collecting private citizens' information from technology providers, we must ensure that safeguards are in place to protect Americans' rights."

The logic behind the Big Brother technology is simple: knowing their customers better allows ads specific to the customer to be targeted to the home. Google operates on the same principle, reading all your email in order to deliver targeted advertising. But the technology in the new generation of DVRs goes so far as to actually film people in the house and record their conversations.

"Devices would utilize technology such as infrared cameras and microphones embedded in DVRs and cable boxes. A patent application filed with the US Patent and Trademark Office by Verizon notes the technology could detect a range of viewer activities. According to the patent application, the set-top device will be able to distinguish 'ambient action … of eating, exercising, laughing, reading, sleeping, talking, humming, cleaning' and more'."

What do they mean by "more"? Well, just imagine a bedroom:

"The application further notes that ambient action also 'comprises an interaction between the user and another user' which are described as 'at least one of cuddling, fighting, participating in a game or sporting event, and talking'. That information would then be used to deliver targeted ads to your living room. Other companies, such as Intel and Comcast have explored similar technology.

"Although this DVR technology is in its conceptual stages," warn congressmen Capuano and Jones, "it is important that Congress establish clear boundaries before it becomes reality. Right now, there is nothing preventing companies from utilizing the technology, no obligation to notify the consumer before it is used and no obligation to give consumers the chance to opt out. Too often, Congress is far behind when it comes to advancements in technology and is forced to update regulations that are decades old so that they are meaningful for today's innovations. The "We Are Watching You Act" does not prohibit companies from developing this technology. It simply lets consumers make their own decisions about whether or not it belongs in their homes.

"If someone is watching a favorite program while cleaning the kitchen, ads for products that will make appliances shine may show up. If a couple is talking about where to dine over the weekend, or enjoying a crunchy snack while watching the Stanley Cup finals, their television may serve up food suggestions. The patent application even goes so far as to suggest that if a couple is fighting, an advertisement associated with relationship counseling might be selected for them."

It would require only a mild tweak of the software and the unit could call police based on the level of aggression in voices and the proximity of the two antagonists to each other. You can argue, at one level, that this might be a good thing, it might save lives. Indeed so, but it brings us back to the fundamental question posed at the start of this book: how much control do we ultimately wish to give the government over our lives?

The proposed "We Are Watching You Act" – if it ever passes into law – would require the device to warn consumers:

"When the recording device is in use, the words WE ARE WATCH-ING YOU would appear, large enough to be readable from a distance, for as long as the device is recording the viewing area. If consumers opt out of the new technology, companies are required to offer a video service that does not collect this information but is otherwise identical in all respects."

This is 1984, 2014-style, albeit initially at the behest of commercial giants rather than the state. Don't be lulled into a false sense of security by that, however. As you will see shortly, the State prefers to work through secret partnerships with big multinational companies.

One of the first devices on the market to test the privacy limits is the Kinect, a new camera built into the latest Microsoft X-box gaming consoles. The Kinect is designed to provide real interactivity by tracking gamers' movements and voice commands, but in doing so it opens the door 24/7 to your living room or bedroom, or your child's bedroom.

"[It] gets a perfect view of your living room. It's always listening for voice commands, even when you turn the Xbox off. It can even read your heartbeat with the right software," reports *The Verge*.[44]

Amid revelations that Microsoft has handed over private data to government agencies in the NSA surveillance scandal, many no longer trust the computing giant's loyalties. Perhaps with good reason. We'll return to that later. The take home point is that householders are being bombarded with so many security breaches simultaneously that it is impossible to

---

44 "Could the NSA use Microsoft's Xbox One to spy on you?" by Sean Hollister, The Verge, 16 July 2013, http://www.theverge.com/2013/7/16/4526770/will-the-nsa-use-the-xbox-one-to-spy-on-your-family

catch them all. If it's not the X-box it's the laptop. If it's not the laptop, it's your phone. If it's not the phone, it's your smart meter. And so on it goes. People are literally being overwhelmed, and the State is piggy-backing on commercial service providers to do it. It is insidious, but effective.

Then there are the motives of others in the commercial sector to consider. In the same way that insurance companies get you to sign waivers so they can access records that medical clinics or other insurers hold on you, the same will become true of smart meter data. The National Institute of Standards and Technology (NIST) identified insurers as one of 17 groups expected to make use of smart meter data "to determine health care premiums based on unusual behaviors that might indicate illness".[45]

The mere fact that a law exists saying you cannot be "compelled" to provide such data, has no impact on the Law of Contracts where a person *agrees* to surrender their privacy in return for some perceived contractual gain (provision of insurance services). Likewise when you sign up for X-box or a range of other services, the fine print of the contracts will see you signing away your ultimate privacy rights. If you refuse to sign, you won't get the service.

Temptation over new toys and tech is a huge motivator. That, coupled with the "it won't happen to me" syndrome. We all believe that national security and surveillance issues happen "over there" to "someone else", someone obviously more deserving of the attention. Tell that to "spread 'em and cough" victims. What we don't anticipate in the heat of the moment is that the freedoms we give away today are effectively given away forever, and that the common enemy we perceive today may not be the surveillance target five years from now when the system has become so large, and the protections have been knocked aside, that it can turn its attention to those who dissent.

For the past four hundred years, individual nation-states have provided a competitive market for ideas and freedoms to flourish. The fact that no single state was capable of knocking out all the others , and turning into a Death-Star type empire, meant that dissenters in one country always had another they could flee to, out of reach of those who would persecute or crush them. The accumulation of powers in a new global regime of the "international community" today removes those fundamental protections. Such a regime cannot take effect until the infrastructure to

---

45 http://spectrum.ieee.org/energy/the-smarter-grid/privacy-on-the-smart-grid

support global governance is firmly in place and the international treaties giving them power are locked and loaded. What we are currently seeing is both of those preliminaries rapidly being finalised: treaty negotiations are underway, and control technology is being standardised around the world – both of these things under the supposed umbrella of combating "climate change".

The prospects for control of vast numbers of people are significant. As the Cuijpers report on the smart meters concludes, our homes are swiftly losing what little resemblance they formerly had to castles:[46]

"The house is rapidly losing its character as privacy's fortress, with directional microphones recording in-house conversations, cameras seeing through walls, thermal imagers detecting heat emissions, household appliances incorporated in the Internet of Things, the home computer permanently connected to the Internet, and private information such as personal texts, photos, books and music no longer stored in desks or on shelves but instead in the cloud. Smart meters are yet another addition to this increasing transparency of the home. This requires careful consideration of the cumulative effect of the various developments that allow insight into how people live, in the one place where people most of all must feel free to do what they like. If our home will no longer be our castle, the house may be energy-efficient but it will be a cold place to live."

If you haven't already guessed, smart grids and smart meters are an implementation of the United Nations Agenda 21 programme. Agenda 21 was the creature of the 1992 Earth Summit at Rio, a list of rules designed to be put in place by every national government and local authority, to change the way we live.

Supporters of the smart meters deny this, saying there is nothing in the text of Agenda 21 mentioning smart meters or grids. They're right – to a point. The Agenda 21 document was drafted back in 1992. Smart meters were not available. Even the internet as we know it didn't really exist back then. It was a scoping document discussing principles, and leaving open the possibility that new technologies would develop that could help implement Agenda 21 principles. The concept behind smart meters is clearly signalled in Agenda 21:[47]

---

46 Cuijpers & Koops (2012), 'Smart metering and Privacy in Europe', in: *European Data Protection: Coming of Age*, Dordrecht: Springer, p. 269-293
47 Clause 4.24, Agenda 21, http://sustainabledevelopment.un.org/content/documents/Agenda21. pdf

"Without the stimulus of prices and market signals that make clear to producers and consumers the environmental costs of the consumption of energy, materials and natural resources and the generation of wastes, significant changes in consumption and production patterns seem unlikely to occur in the near future."

At Clause 6.41, subclause i(iv), the UN calls for "the introduction of environmentally sound technologies within the industry and energy sectors". At Clause 7.69(c), all countries are instructed to "adopt standards and other regulatory measures which promote the increased use of energy-efficient designs and technologies and sustainable utilization of natural resources in an economically and environmentally appropriate way."

At Clause 9.12(c) governments are instructed to "promote the research, development, transfer and use of improved energy-efficient technologies and practices, including endogenous technologies in all relevant sectors, giving special attention to the rehabilitation and modernization of power systems."

You can anchor the development of mercury-filled CFL lightbulbs and smart meters to all these and many more citations in Agenda 21. Don't let anyone fool you with the line that because these technologies are not specifically-named in a 1992 document, they are not part of the plan. The word "electricity" is not mentioned in Agenda 21 either, but anyone with a windfarm on their doorstep knows full well where the initiative came from.

Still, for those of you who think of Nanny State surveillance less in terms of *The Bourne Identity* and more in terms of *Get Smart*, there is some corroborative support for your position. A recent security bulletin alerted customers of a "smart" Japanese toilet system now being marketed around the world that they could get bogged down in the device because of hackers.

The Satis automated toilet features every whizz-bang known to mankind and can be controlled through a smart phone app (meaning you don't have to actually touch the toilet itself). The only problem is, the app used the same security code for all users, meaning anyone who downloaded the app could control any Satis toilet, anywhere, with devastating results:[48]

"Attackers could cause the unit to unexpectedly open/close the lid, activate bidet or air-dry functions causing discomfort or distress to user."

---

48 https://www.trustwave.com/spiderlabs/advisories/TWSL2013-020.txt

Not to mention suddenly flushing the bidet unit underneath an intended victim.

Remember, while the push towards smart grids is being sold to consumers as a matter of convenience, cost-cutting and home automation, the real reason governments are swinging in behind is the data-mining, the remote control capacity and the supposed connection to climate change.

Smart grids on their own, however, are merely one sticky strand in the technology web now spread out across the flightpath of your life. If you live in Europe, South East Asia or the English-speaking countries, chances are you are well and truly caught in the web, as you are about to see.

CHAPTER FOUR

The Falcon And The Snowden

"They're just as paranoid and dangerous as we are. I don't know why I ever thought any differently."
– Falcon And The Snowman

In the 1985 movie *The Falcon & The Snowman*, we learn the true story of how two young American men – one of them working as a civilian contractor in a top secret intelligence gathering facility – become disillusioned about the duplicity of their Government and its spying operations. One cable that came through, for example, detailed a CIA plot to oust Australian Prime Minister Gough Whitlam in the 1970s. The characters decide to sell secret information to the Soviets.

In May 2013, history repeated itself to an extent, when world newspapers broke the story of 29 year old CIA whistleblower Edward Snowden. Snowden, who'd worked for both the CIA and the NSA before joining civilian contractor Booz Allen Hamilton, decided to quit his job in spectacular style, by blowing the lid off Western intelligence surveillance operations.

"The NSA has built an infrastructure that allows it to intercept almost everything," Snowden told Glenn Greenwald and Ewan Macaskill of the *Guardian*. "With this capability, the vast majority of human communications are automatically ingested without targeting. If I wanted to see your emails or your wife's phone, all I have to do is use intercepts. I can get your emails, passwords, phone records, credit cards.

"I don't want to live in a society that does these sort of things … I do not want to live in a world where everything I do and say is recorded. That is not something I am willing to support or live under."

The track towards this disclosure had taken decades. In the 1995 book *The Paradise Conspiracy*, details of a top secret agreement known as the UKUSA Treaty were disclosed.[49]

"It is interesting to note some of the behind the scenes machinations that accompanied my Official Information Act requests on these matters. A week before my first request to Defence I had fired in an OIA application about the UKUSA Treaty, a top secret military intelligence treaty that New Zealand was a signatory to. That request had resulted in a call to TV3 news director Rod Pedersen's office from defence officials representing the Minister of Defence, Warren Cooper.

"Pedersen's executive assistant, Carol van Stockum, told me they'd asked for an urgent meeting with him in Cooper's Beehive office, and they were prepared to fly Pedersen down to Wellington at their expense. Rod and Carol stalled for time – eventually the follow-up calls dwindled and stopped.

"Thus, when I faxed off my first request on the SAS operations the following Friday, Defence chiefs had already been rattled. On Friday afternoon I got a phone call from a businessman I'd met through Spook, a man he claimed had intelligence links – although the person concerned denied it vehemently. However, our mutual friend gave me a warning on this occasion.

" 'Your Official Information request is creating waves. This is a dangerous business, so I want you to be extremely careful and vigilant around your house and car.[50] You'll get a letter on Monday denying the existence of the information you're seeking. I wish we'd had a chance to discuss this before you put the request in, but what's done is done.'

"The UKUSA Treaty is classified on the 'If I told you, I'd have to kill you!' level. Its existence is neither confirmed nor denied, but essentially it relates to interception of radio and phone communications. Virtually all international calls, cellular calls and radio transmissions can be and are intercepted – those containing certain keywords are captured by computer for further analysis.

---

49 *The Paradise Conspiracy* by Ian Wishart, 1995, p61
50 In the final stages of preparing *The Paradise Conspiracy* for release, my car was sabotaged by someone who'd managed to bypass the sensitive car alarm and inject silicon into the brake fluid pipes, causing brake failure on the motorway. A police investigation confirmed the sabotage. No culprits were ever found. See "Postscript", p327 of the book.

"I could feel the goosebumps rising as my contact spoke. What made it even more chilling was the fact that on Monday, as predicted, a letter turned up from the Ministry of Defence turned up denying the existence of the information I was seeking. It was the fastest reaction to an Official Information Act request that I'd ever had."

The UKUSA revelations in the book, seemingly minor as they were, were fleshed out significantly by researcher Nicky Hager for his 1996 book "Secret Power", which gave the UKUSA listening apparatus a codename: Echelon. This was a series of listening posts in America and around the world in the UK, Canada, Australia and New Zealand. These five countries all agreed to spy on the telephone calls of the citizens of other countries, in order to circumvent national restrictions on spying on your own citizens. For example, the USA might eavesdrop on New Zealand and Australia, while the ANZACs would return the favour and spy on the Americans and the Brits. The relevant information was sifted and shared.

Fast forward to Edward Snowden's revelations in 2013 and the spooks have come a long way since the primitive surveillance systems of the 1990s.

"You are not even aware of what is possible," Snowden told the *Guardian*. "The extent of their capabilities is horrifying. We can plant bugs in machines. Once you go on the network, I can identify your machine. You will never be safe whatever protections you put in place."[51]

That doesn't stop Snowden from trying some "basic" protections regardless:

"He is deeply worried about being spied on," writes the *Guardian* after speaking to him in a Hong Kong hotel room. "He lines the door of his hotel room with pillows to prevent eavesdropping. He puts a large red hood over his head and laptop when entering his passwords to prevent any hidden cameras from detecting them."

Even then, Snowden realises it is probably futile. No red hood is going to save him from the particular variety of big bad wolf that wants a piece of him.

Anything, then, on your computer, your cellphone is free to be read by the Government. It wasn't always the case. In 1998, renegade British MI6 agent Richard Tomlinson came to New Zealand looking for a tell-all book publishing deal about MI6 involvement in the death of Princess Diana. Howling At The Moon Publishing Ltd in Auckland contacted Fourth

51 "Edward Snowden: the whistleblower behind the NSA surveillance revelations", The Guardian, 10 June 2013, http://www.theguardian.com/world/2013/jun/09/edward-snowden-nsa-whistleblower-surveillance

Estate Publishing Ltd in London, using PGP encryption to prevent any Echelon surveillance of the email contents.

Unable to break into the email, British Special Branch Police instead broke into Fourth Estate Publishing's London office, threw staff up against the wall and confiscated the computer the email had been sent to. Back in Auckland, the New Zealand publishing company was slapped with a gagging writ.

So, when Edward Snowden says he believes his life is now in danger, there's no particularly good reason to assume he is wrong:

"Yes, I could be rendered by the CIA. I could have people come after me. Or any of the third-party partners. They work closely with a number of other nations. Or they could pay off the Triads. Any of their agents or assets."

He had a life. A $200,000 job, a girlfriend. Now he's hunted.

"I'm willing to sacrifice all of that because I can't in good conscience allow the US government to destroy privacy, internet freedom and basic liberties for people around the world with this massive surveillance machine they're secretly building."

So what does the 'machine' do today, that it could not back in 1996?

One document released by Snowden, "XKeyscore"[52] is dated 2008. It is slugged "top secret" on its front page for the attention of "USA, AUS, CAN, GBR, NZL" – the same five parties who signed the 1947 UKUSA Treaty that created Echelon. In intelligence terms, these five countries control the world with their access to the most sophisticated intel systems on the planet.

XKeyscore, says the briefing, keeps a "rolling buffer" of between three and thirty days worth of intercepted emails, web-searches and other communications information, which can be "mined" for relevant keywords. Users at 150 mostly secret locations around the world can log in to any of the system's 700 servers and search through the system for information on anyone.

When it hones in on a 'target' through their phone number, IP address, email or other trackable activity, it quickly sucks your computer dry of all its cookies and passwords, chat conversations, webmail data, so that someone in an office somewhere can, in effect, be you.

Supposing you want a list of everyone on the planet who, in the past

52 http://www.theguardian.com/world/interactive/2013/jul/31/nsa-xkeyscore-program-full-presentation

24 hours, has searched for the phrase 'Echelon' on Google. No problem. Within minutes vast stacks of computer memory have been utilised to bring you the list of 3,521 people who searched that word. From there it's simply a case of filtering the data further until you reach a list of "likely suspects".

One of the advantages of XKeyscore is that it's programmed to pick up "chatter" – the flurry of internet traffic that sometimes precedes terror attacks, or which reflects an upsurge of activity on a particular site. The machine rings alarm bells and says, 'hey, take a look at this'.

Another thing XKeyscore is programmed to look for is people using encryption on the internet. It turns out to be a little like Frodo slipping on the ring in *Lord of the Rings*; suddenly, the dark lord Sauron can see everything he's up to, even though he's invisible to everyone else. Little wonder the Fourth Estate Publishing Company got raided after receiving an encrypted email.

The briefing document suggests system operators pose questions like "show me all PGP usage in Iran" as an example of how quickly the system can round up possible suspects online.

Many families and businesses use VPNs, or virtual private networks, to communicate securely between homes and offices. Don't worry, XKeyscore can see right through those walls as well. Need to find a German-speaking target in Pakistan? No problem there either, as XKeyscore documents internet searches by the language of the user then compares them to the language of the location.

The system is so integrated with Google that it can, for example, reverse look-up a Google Maps query you made 29 days ago, and from that simple query find you and locate where you are at this instant. If you are an al Qa'ida operative, that's one of the techniques they use to put a cruise missile into your camel hut.

The briefing document discloses a question: "I have a jihadist document that has been passed around through numerous people, who wrote this, where were they?" The answer has been blacked out by Snowden, presumably on the grounds it would compromise security.

As of 2008, the document boasts more than 300 terrorists had been captured by XKeyscore, although specific examples were also blacked out.

The thing to remember is that XKeyscore is not 'the Machine', it is merely one part of 'the Machine'. There are literally dozens of interlocking systems being used. When it finds a person of interest, all of those interlocking systems coalesce into a complete investigative process.

In Chapter One, we talked a little about conspiracy theory versus coincidence theory, and random events that key players were able to take advantage of. One of those seemingly random events was the attack on the World Trade Centre in New York by al Qa'ida.

There is plenty of evidence that al Qa'ida operatives carried out the attacks. Osama bin Laden admitted it, and Arabic-speaking men copying airliner flight manuals had been seen in New Zealand three months prior to the twin towers impacts:

"New Zealand's Security Intelligence Service (SIS) is investigating a visit by two men believed to be Arabs to a photocopying centre months before the terrorism attacks in the United States, the *New Zealand Herald* reported.[53]

"The shop owners in Hamilton said the men made copies of an aircraft manual in Arabic. The owners noticed the numbers 757 and 767 and drawings of planes and instruments.

"The hijacked airliners flown into the World Trade Centre and Pentagon on September 11 were a Boeing 757 and a 767."

Does the fact that al Qa'ida staged these attacks mean, however, that others have not taken advantage of them? One of the themes of this book is "convergence of opportunities", which is where things come together and become game changers. In a way that's a good analogy. The bounce of the football can make or break a match at a crucial moment. So too in history.

In 1941, one of President Roosevelt's trusted advisors, a special envoy on a mission to Japan, sent word to the President that Japan planned to attack the US naval base at Pearl Harbor, Hawaii.[54]

Roosevelt chose to sit on the information, knowing that such an attack would inevitably drag an unwilling US public into the war. Until such time as an attack happened, Roosevelt knew he had no moral authority to pre-empt. So he sat, twiddling his thumbs.

He later described it as a "date which will live in infamy".

In the same fashion, there are those who suspect elements within the US government knew in advance about the 9/11 attacks.

---

53 "Arab photocopy link to terror", TVNZ News, 18 October 2001, http://tvnz.co.nz/content/62635/423466/article.html
54 I know this because I have seen the letters. The special envoy was the father of a friend of my family's.

Certainly there is evidence that fairly explicit warnings fell on deaf ears. Like the FBI's Phoenix memo, which warned about a possible airliner strike on the trade centre. FBI Director Robert Mueller admitted as much at the Senate Judiciary Committee hearing in 2002:

"Of the warnings that we had – the stopping of Moussaoui, the arrest of Moussaoui, brought the bureau, and particularly the agent in Minneapolis, to the belief that this individual is the type of individual that could and might be the type of individual to take a plane and hijack it. And in fact, if I'm not mistaken, in one of the notes, the agent in Minneapolis mentioned the possibility of Moussaoui being that type of person that could fly something into the World Trade Center."[55]

Before 9/11 happened, the North American air defence command NORAD had run crisis scenarios involving airliners used as weapons of mass destruction:[56]

"In the two years before the Sept. 11 attacks, the North American Aerospace Defense Command conducted exercises simulating what the White House says was unimaginable at the time: hijacked airliners used as weapons to crash into targets and cause mass casualties.

"One of the imagined targets was the World Trade Center. In another exercise, jets performed a mock shootdown over the Atlantic Ocean of a jet supposedly laden with chemical poisons headed toward a target in the United States. In a third scenario, the target was the Pentagon – but that drill was not run after Defense officials said it was unrealistic, NORAD and Defense officials say."

Those who see method in all madness cite this as proof the government knew in advance, but if they did know in advance, why bother preparing for an event they had no intention of preventing? Read another way, and with an already known Middle Eastern trend of using vehicles packed with explosives driven by suicide bombers directly into buildings, and a previous attempt to blow up the World Trade Centre, it wasn't a major leap up the pay-grade to correctly assume someone might try using planes the same way.[57]

55 "Who Will Be the Red Flag Scapegoats?", ABC News, 18 May 2002, http://abcnews.go.com/US/story?id=91624&page=4
56 "NORAD had drills of jets as weapons", USA Today, 18 April 2004, http://usatoday30.usatoday.com/news/washington/2004-04-18-norad_x.htm
57 I have written elsewhere of this, but in August 2001 I was commissioned to write a pilot episode for a proposed new Canadian/NZ/Australian TV production. My pilot script was entitled "The Two Towers" and detailed a simultaneous strike by al Qa'ida militants on the Auckland Skytower and Sydney Centrepoint Tower. The script was due on a Canadian producer's desk on 10 September

When trying to pigeonhole something in the bureaucratic sphere as either conspiracy or coincidence, there is one other c-word variation that can often be applied to government agencies: cock-up theory.

Bureaucracy, by its very nature, involves vast amounts of paper shuffling, memoranda, meetings, lists, meetings about meetings. In many cases bureaucracy is a make-work scheme for people who might not otherwise find paid employment in the real world. Despite the glamourisation of Hollywood, even the intelligence and law enforcement agencies are prone to the worst excesses of the public service everywhere.

That means that sometimes, even though "conspiracy" might be a tempting description, "cock-up" fits the facts even better. 9/11 was a cock-up of epic and tragic proportions, and it quickly descended into a butt-covering exercise at all levels.

John Farmer was a senior counsel to the 9/11 Commission, and in his book about the Commission investigation he writes they "discovered that what had occurred that morning – that is, what government and military officials had told Congress, the Commission, the media, and the public about who knew what when – was almost entirely, and inexplicably, untrue."[58]

They lied, he writes, to protect their backsides and make it look like their own agencies were more in control than the record suggests.

The memos, emails and testimony to the 9/11 Commission tell a story. They show that the NSA's worldwide snooping operation was picking up reports of unspecified threats against Western interests. National Security coordinator Richard Clarke briefed Condoleeza Rice on 23 March 2001 that terrorists may use a suicide truck bomb as their "weapon of choice" to blow up the West Wing of the White House.[59]

On 19 April, Clarke advised "Bin Laden planning multiple operations", but the Americans didn't know where, and assumed it was their embassies and military bases overseas.

"In May 2001," reported the Commission, "the drumbeat of reporting grew louder with reports to top officials that "Bin Laden public profile

---

2001, and it was. If the idea could occur to a writer on the other side of the planet, it could equally occur to defence planners and FBI field agents. There is a big difference however between an idea of how something could be done, and specific information on when it was actually being done.
58 "The Ground Truth", by John Farmer, September 2009, http://www.stltoday.com/entertainment/books-and-literature/reviews/the-ground-truth/article_97bbba18-d273-5995-96ed-33a05a57e1ed.html
59 9/11 Commission Report, p255, http://www.9-11commission.gov/report/911Report.pdf

may presage attack" and "Bin Laden network's plans advancing." In early May, a walk-in to the FBI claimed there was a plan to launch attacks on London, Boston, and New York. Attorney General John Ashcroft was briefed by the CIA on May 15 regarding al Qaeda generally and the current threat reporting specifically. The next day brought a report that a phone call to a U.S. embassy had warned that Bin Laden supporters were planning an attack in the United States using "high explosives." On May 17, based on the previous day's report, the first item on the CSG's agenda was "UBL: Operation Planned in U.S." The anonymous caller's tip could not be corroborated."

By late June, the intelligence chatter was suggesting attacks somewhere would be "spectacular", and Osama bin Laden himself appeared on Arab TV warning that the next few weeks would contain "surprises". Al Qa'ida released a new recruitment video.

"Clarke," says the Commission report, "wrote that this was all too sophisticated to be merely a psychological operation to keep the United States on edge, and the CIA agreed. The intelligence reporting consistently described the upcoming attacks as occurring on a calamitous level, indicating that they would cause the world to be in turmoil and that they would consist of possible multiple—but not necessarily simultaneous—attacks.

"On June 28, Clarke wrote Rice that the pattern of al Qaeda activity indicating attack planning over the past six weeks "had reached a crescendo."

"A series of new reports continue to convince me and analysts at State, CIA, DIA [Defense Intelligence Agency], and NSA that a major terrorist attack or series of attacks is likely in July," he noted. One al Qaeda intelligence report warned that something "very, very, very, very" big was about to happen, and most of Bin Laden's network was reportedly anticipating the attack. In late June, the CIA ordered all its station chiefs to share information on al Qaeda with their host governments and to push for immediate disruptions of cells.

"The headline of a June 30 briefing to top officials was stark: "Bin Laden Planning High-Profile Attacks." The report stated that Bin Laden operatives expected near-term attacks to have dramatic consequences of catastrophic proportions."

It's a big report, at some 585 pages. The point is, US intelligence knew something was coming, but they didn't know precisely what was coming. Among the items of evidence pointing to cock-up theory was the revelation that the CIA concentrated its attention in a regionally-focused fashion.

If an al-Qa'ida man popped up in Malaysia, the CIA's South-East Asia desk were all over it like a rash – until he left their specified region. In contrast, said the 9/11 Commission, the FBI tended to operate on a "total case" basis with a controlling officer on it from beginning to end. The FBI's focus was continuity, the CIA's was zonal. The two agencies didn't really understand how each functioned. Al Qa'ida fell between the cracks.

"When the trail went cold after the Kuala Lumpur meeting in January 2000, the [CIA] desk officer moved on to different things. By the time the March 2000 cable arrived with information that one of the travelers had flown to Los Angeles, the case officer was no longer responsible for follow-up. While several individuals at the Bin Laden unit opened the cable when it arrived in March 2000, no action was taken."[60]

The point to remember here is that this was Edward Snowden's XKeyscore, or its predecessor, in action. Even despite having the most sophisticated eavesdropping network in the world, the data was useless if humans were not processing it properly.

Having said all that, the 9/11 disaster played right into the hands of control freaks everywhere. Suddenly, every plane passenger was a suspect. Everyone with olive skin was a suspect. How could security be so lax that halfwits with box cutters could get away with such an outrage, we all wondered?

In the urge to feel safe travelling again, the public willingly gave up more of their freedoms. As indicated, the evidence that al Qa'ida carried out the attacks using real hijacked planes is overwhelming. However, the unanswered question is whether someone incited them into it.

One of the draconian new laws enacted to ostensibly deal with the terrorist threat was the Patriot Act.[61] This piece of legislation passed into law on October 26 2001, just three days after its introduction.

The Patriot Act itself was an amalgam of a raft of earlier bills introduced to Congress within three weeks of 9/11. The significant thing about all of this is just how fast lawmakers were able to draft new laws amending a clutch of other laws very precisely. If the CIA and FBI had missed the boat on the terror warnings, legislators were surprisingly ready with exactly the powers that the President was demanding.

---

60 Ibid, p268
61 Technically the USA PATRIOT Act, the capitalisation being an anagram of Uniting and Strengthening America by Providing Appropriate Tools Required to Intercept and Obstruct Terrorism (USA PATRIOT ACT) Act of 2001

The rapid introduction of a swathe of anti-terror laws capitalised on the sense of fear the 9/11 attacks had engendered, it was a political version of "shock and awe". Among the existing laws amended by the Patriot Act were:

- Electronic Communications Privacy Act
- Computer Fraud and Abuse Act
- Foreign Intelligence Surveillance Act
- Family Educational Rights and Privacy Act
- Money Laundering Control Act
- Bank Secrecy Act
- Right to Financial Privacy Act
- Fair Credit Reporting Act
- Immigration and Nationality Act
- Victims of Crime Act of 1984
- Telemarketing and Consumer Fraud and Abuse Prevention Act

The essence of Patriot was to make it much easier for intelligence and law enforcement agencies to investigate suspected terrorists without being caught up in red tape. Some of the other laws, like the Foreign Intelligence Surveillance Act, had been introduced after the CIA scandals of the 1970s and the Watergate break-ins. It had emerged in Congressional hearings in the seventies that elements of the CIA had funded black operations (off the radar, not on the official books) by means of domestic crime like bank robberies, drug trafficking and even murder.[62]

The Patriot Act effectively wiped those sins clean, and turbocharged the eavesdropping system run with the other "Five Eyes" powers – New Zealand, Australia, Canada and Britain.

Among the powers given to agencies were the right to detain immigrants indefinitely, the right to search homes or offices without the knowledge or consent of the owners or occupiers, and massively extended rights to search emails, phone records, business records and all those things now being trawled on the internet.

Parallel developments led to the creation of the Department for Homeland Security, with upgraded powers for the TSA at airports (of the 'spread em and cough' variety).

Everywhere you looked, George Orwell's 1984 was increasingly the 'new

---

62 See The Crimes of Patriots by Jonathan Kwitny, Simon & Schuster NY, 1987

normal' as the public resigned themselves to new restrictions. But just how successful were the new laws at actually preventing terror attacks?

In a ten year review of the Patriot Act in 2011, journalist Peter Huck wrote:

"There is no evidence – or at least none put forward by US authorities – that the Patriot Act has foiled a single terrorist plot.[63]

"Out of the 192,499 'national security letters' used by the FBI to sift through phone, computer and credit records without judicial oversight or probable cause, only one resulted in a terror-related conviction. The ACLU [American Civil Liberties Union] says this conviction could have been secured using other laws.

"Of the 143,074 letters issued between 2003 and 2005, 53 led to reported prosecutions. Seventeen were for money-laundering, 17 for immigration and 19 for fraud. None was related to terrorism, suggesting national security letters are used as a fishing expedition, snaring criminals who might be prosecuted under other laws."

If you think that's strange, consider the case of Adam McGaughey, a fan of the TV series *Stargate SG1*. He was charged, under the Patriot Act, with copyright infringement of the programme on his fan website.[64] He had nothing to do with terrorism or terrorists, but the law changes introduced by the Patriot Act gave authorities the chance to prosecute.

The *New York Times* reported that within just two years of being passed, the Patriot Act had become the default setting for all kinds of criminal investigations:[65]

"The Bush administration, which calls the USA Patriot Act perhaps its most essential tool in fighting terrorists, has begun using the law with increasing frequency in many criminal investigations that have little or no connection to terrorism.

"The government is using its expanded authority under the far-reaching law to investigate suspected drug traffickers, white-collar criminals, black-mailers, child pornographers, money launderers, spies and even corrupt foreign leaders, federal officials said.

"Justice Department officials say they are simply using all the tools

63 "Protecting liberty by taking it away", by Peter Huck, NZ Herald, 5 November 2011, http://www.nzherald.co.nz/world/news/article.cfm?c_id=2&objectid=10764033
64 "MPAA Uses Patriot Act to Enforce Copyright Law (VANITY)", FreeRepublic.com, 30 March 2004
65 "U.S. Uses Terror Law to Pursue Crimes From Drugs to Swindling", New York Times, 28 September 2003, http://www.nytimes.com/2003/09/28/politics/28LEGA.html

now available to them to pursue criminals – terrorists or otherwise. But critics of the administration's antiterrorism tactics assert that such use of the law is evidence the administration is using terrorism as a guise to pursue a broader law enforcement agenda."

You can probably make a good case that catching a criminal is catching a criminal – a worthy end in and of itself. The flip side, however, is that society has never chosen to delegate this level of power to governments before, and while the laws can be used to good effect they can also be misused. The 'fact' that they are now also targeting criminals is not the issue – it is the 'fact' that they can now target virtually anyone, even you, that should be a concern.

Twelve years down the track, even Jim Sensenbrenner, the US senator who authored the Patriot Act, now believes the US Government is massively overstepping its powers:[66]

"As the author of the Patriot Act, I am extremely troubled by the FBI's interpretation of this legislation. While I believe the Patriot Act appropriately balanced national security concerns and civil rights, I have always worried about potential abuses. The Bureau's broad application for phone records was made under the so-called business records provision of the Act. I do not believe the broadly drafted FISA order is consistent with the requirements of the Patriot Act. Seizing phone records of millions of innocent people is excessive and un-American."

It is now clear that the balance of power between citizen and state has tipped, perhaps irreversibly – barring a revolution.

As always, the slide to a kind of totalitarianism is based on the argument "we know best". But it is not just surveillance and military adventures that this reasoning applies to. Locked in to the mix is the increasing reliance on globalisation as a concept, and standardisation of rules and laws across the planet.

These laws are almost non-existent against global corporations, but heavily oppressive on smaller regional businesses trying to compete. Did you know, for example, that there's a war on over what you should be allowed to eat, what vitamin supplements you should be allowed to take and how much you should be allowed to know about what's actually in your food?

---

66 Letter from Jim Sensenbrenner, 6 June 2013 http://sensenbrenner.house.gov/UploadedFiles/Sensenbrenner_Letter_to_Attorney_General_Eric_Holder.pdf

CHAPTER 5

Health Control

"Recent research funded by the pharmaceutical industry
and the accompanying flood of new drugs for a range
of ills threaten to 'medicalize' every human condition
and behavior. Now, childhood shyness and internet
browsing could be reclassified as mental disorders
under controversial new guidelines, warn experts..."
– News, Feb 2012[67]

Medical science. It is based on "science", right? You could be forgiven
for wondering as you read this next chapter. Worldwide, the health sector is worth trillions in the global economy.[68] Go back a century and a
half, before the widespread use and development of antibiotics and other
modern medical tools, and life was often short and brutish. There is no
question that we live longer now – the ultimate test of a health system.

However, and with a big 'H', all is not necessarily as it seems.

The trend throughout the past century has been towards developing
unique, human-designed drugs and medicines to treat disease, fostered
by an inherent belief that if humans throw enough money and resource
at something they can find a solution to any problem.

Unique medicines are expensive to research, and big pharmaceutical

67 http://worldtruthtoday.wordpress.com/2012/02/13/shyness-and-internet-browsing-may-soon-
be-classified-as-mental-illness/
68 Forecast to reach $3 trillion per annum by 2015, http://www.reportlinker.com/ci02241/Healthcare.html

companies need a return on their massive scientific investments. Accordingly, the drugs are priced high. Chemotherapy drugs for cancer patients, for example, can run to US$11,000 or more per two ml dose.

The chemical companies behind these drugs are massive multinationals, but they didn't start off big. GlaxoSmithKline (GSK), for example, began as baby milk powder company Glaxo in the tiny village of Bunnythorpe in New Zealand, in 1904. It expanded into the UK and America, merging with other pharmaceutical companies as it went, until today it has a market capitalisation in Britain of around 75 billion pounds. The story of the growth of massive multinational corporations is closely related to the growth of globalisation.

There is a good reason for mentioning GSK. In 2012, the book *Vitamin D: Is This The Miracle Vitamin?* was launched and promoted around the world. The book took a look at the peer-reviewed scientific studies on vitamin D and concluded that a large number of modern health problems appear to be directly related to low levels of vitamin D.

Studies, for example, show a massive reduction in cancer mortality in people who have high vitamin D levels, and a similar effect is seen for heart disease. One study showed women diagnosed with breast cancer whose vitamin D levels were low were 90% more likely to see their cancer spread and 73% more likely to be dead at the end of 10 years.[69]

When some of these studies were detailed in radio interviews to promote the Vitamin D book, they resulted in a complaint to New Zealand's largest radio network, alleging the vitamin D claims were "quackery". The radio station passed on the anonymous complaint to the publisher, and the publishing company traced the hotmail email address of the sender back to a named individual. The named individual turned out to be one of GSK's top managers in New Zealand.

Why, you might wonder, would the world's fourth largest pharmaceutical company be remotely interested in attacking a popular book on vitamin D?

Well, it turns out GSK is a big manufacturer of chemotherapy and other cancer treatments, like Tykerb for breast cancer. If the vitamin D scientific studies are correct, and the incidence of cancer in the community can be cut by even just 30% by boosting vitamin D intake, then cancer drug companies like GSK face financial ruin.

---

69 http://www.thestar.com/life/health_wellness/2010/03/05/breast_cancer_survivor_credits_vitamin_d_for_recovery.html

It's been said before in this book that things are not usually black and white. This is another case in point. Whilst the benefits to the community from reducing the incidence of cancer are massive, events that would bring pharmaceutical research to its knees are not necessarily good for the community.

Under Obamacare, and certainly under other socialised Western health systems like New Zealand's, the cost of pharmaceutical drugs falls largely on taxpayers. A big cut in revenue streams because of a drop in cancer will only see margin prices for pharmaceutical drugs increase even further, to make up for the drop in volume.

Do big pharmaceutical companies actually have a vested financial interest in keeping people sick? There are some who believe so. You can call it Pharma-grade Munchausen's Syndrome by Proxy, in that these companies rely – for their continued business growth – on a steady stream of sick people. If the population gets healthier, demand for their services drops.

There are more clues, however, that big pharmaceutical companies do not have the best interests of the public at heart, no matter how much government health officials insist to the contrary. Increasingly in recent years, investigators have found fraudulent science and even blatant criminality at the heart of many of the so-called scientific tests of major new drugs.

"In the past four years," reports the *Times of India*,[70] "leading members of Big Pharma like GlaxoSmithkline, Pfizer, Johnson & Johnson, Astra-Zeneca, Merck, Abbot, Eli Lilly and Allergen have paid about $13 billion in fines to settle charges of misleading marketing, promising what drugs don't do, bribing doctors to get their drugs prescribed, causing sometimes fatal side-effects, and other crimes. The patients targeted by them ranged from children to dementia-afflicted senior citizens. An analysis of their total revenues and the income from the drugs they are charged with shows that while huge, the fines are at best slaps on the wrist – their jaw-dropping revenues far outweigh the penalties.

"Here are the facts: GlaxoSmithKline was fined $3 billion by the US Justice Department for marketing drugs for unapproved uses, paying kickbacks to doctors and Medicare system, downplaying known risks of certain drugs. They sold Paxil, an antidepressant, to children for whom it was not shown to work. They sold Wellbutrin, another anti-depressant, as

70 "Unfair practices: Pharma companies fined $13bn in 4 years" by Subodh Varma, Times of India, Jul 16, 2012, http://articles.timesofindia.indiatimes.com/2012-07-16/international-business/32697612_1_bextra-avandia-warner-lambert

a pill for weight-loss and erectile dysfunction. They sold the anti-diabetic pill Avandia concealing data that showed it increased cardiac risks. But in the years it took for all this to come through GlaxoSmithkline had made $11.6 billion on Paxil, $5.9 billion on Wellbutrin and $10.4 billion on Avandia. That's $27.3 billion – about 9 times the fine they are paying now to settle investigations.

"Pfizer, the world's biggest pharma company with annual revenue of over $67 billion last year, paid up $2.3 billion in 2009 to settle a similar investigation. The drugs involved were Bextra, Geodon, Zyvox and Lyrica. Pfizer had been using illegal methods to sell them, like giving junkets and cash to sales reps for pushing the anti-arthritic pain killer Bextra as an all-purpose pain killer.

"The reason for Pfizer's huge fine was that it included $1.3 billion for criminal liability – because this was the second time they had been caught. Earlier, in 2004, their subsidiary Warner-Lambert had been fined $430 million for the same violations, and they had promised never to repeat.

"All four of Pfizer's controversial drugs had topped $1 billion in sales before coming under a cloud. And so it goes on. Johnson & Johnson has appealed against an Arkansas judge's ruling to cough up $1.2 billion for off-label marketing of Risperdal, Medicaid fraud and paying kickbacks to nursing care provider Omnicare. But industry experts say that J&J is going to settle with justice department for $2.2 billion and avoid nation-wide penalties which would run into billions. Risperdal is estimated by industry analysts to have earned $24 billion for J&J since it went on sale in 2003 as an antipsychotic drug.

"Abbott Laboratories aggressively pushed the anti-epilepsy blockbuster drug Depakote on elderly dementia patients saying that it helped control their agitation. There was no evidence that it did so. In fact, there was evidence of adverse effects. They also sold it as an anti-schizophrenia drug whereas it was approved only for seizures and bipolar mania. This year, Abbott agreed to settle all claims for $1.6 billion. Abbott had $38.85 billion sales last year."

Many of these big drug companies are heavily involved in "public health" campaigns like vaccinations. But if they are faking the scientific data on commercial prescription drugs, can we trust them on the safety of vaccines? The controversy over the cervical cancer vaccine Gardasil provides some insight.

Gardasil is manufactured by pharmaceutical giant Merck Sharp &

Dohme, a company whose reputation has been seriously stained over the past couple of years, as the *Times of India* reported:

"In 2011, Merck agreed to pay a fine of US$950 million for selling Vioxx, a painkiller for four years before withdrawing it in 2004. It earned about $11 billion from Vioxx, but left behind a trail of patients with heart seizures and strokes."

Again, in 2012, Merck was hit with a US$328 million fine on criminal charges in the US relating to its aggressive marketing of what later turned out to be unsafe drugs.

The question is, as the New Zealand Ministry of Health's main partner in the cervical cancer (HPV) vaccine programme, is the evidence pointing to another debacle over Gardasil?

New Zealand teenager Jasmine Renata died after receiving the vaccine. Official investigations have tried to shift the blame elsewhere, whilst offering no hard evidence of any other cause, but Renata did not die alone. A number of children and young women have died around the world after being vaccinated with Gardasil, and the *British Medical Journal* has recently reported on the case of a 16 year old girl who became infertile after being vaccinated.

After having periods for three years, the teenager stopped menstruating soon after receiving the HPV jab, and has subsequently been diagnosed with "premature ovarian failure". The *British Medical Journal* notes that this condition is very rare in otherwise well teenagers, and it has now been listed as a possible "adverse event" for HPV vaccinations. There was, says the journal, no other rational explanation or known medical cause for her sudden infertility.

Canadian research scientists and Gardasil critics, Doctors Lucija Tomljenovic and Christopher Shaw, have published a series of papers in peer-reviewed medical journals this year casting serious doubt on both the effectiveness and the safety of Gardasil.

They note, first of all, how aggressively Merck has marketed it, in an article in the *Journal of Law, Medicine & Ethics*. They quote the serious debate in medical circles about this vaccine. On the one hand, the American Academy of Pediatrics has welcomed Gardasil and touted its "excellent safety record…a lifesaving vaccine that can protect girls from cervical cancer".

On the other hand, Tomljenovic and Shaw then quote the Association of American Physicians and Surgeons which has stated: "This HPV vaccine costs hundreds of dollars for something that most of the recipients do not even need protection against."

The AAPS also slammed Merck for its marketing, saying that "without adequate testing but with well-placed political funding and lobbyists, Merck pushed for requiring that the HPV vaccine, Gardasil, be given to young schoolgirls as a condition for entering sixth grade. But the disease it supposedly protects against is not even contagious in the school environment."

Does that sound like a pharmaceutical company marketing for genuine health reasons?

If you still want to give Merck the benefit of the doubt, consider this:

"Notably, Merck-sponsored educational programmes delivered by professional medical associations (PMAs) strongly promoting HPV vaccination began in 2006, *more than a year before the clinical trials containing important safety and efficacy data were published*," write Tomljenovic and Shaw.

In other words, long before they knew the drug was safe, Merck were pushing it left right and centre, so much so that in 2006 "Gardasil was named the pharmaceutical 'brand of the year' for building 'a market out of thin air'."

Then there's the question of just how reliable those safety tests were. Instead of pitting the Gardasil injection (based on a potentially neurotoxic aluminium base) against a genuinely neutral placebo, Merck's safety trial matched the injection against an aluminium based placebo. Any nasty reactions caused by the vaccine could be masked by the choice of a potentially problematic placebo in the trial.

"The poor design of existing vaccine safety and efficacy trials may be reflective of the fact that in the past two decades the pharmaceutical industry has gained unprecedented control over the evaluation of its own products," write Tomljenovic and Shaw.

The former Editor-in-Chief of the *New England Journal of Medicine* has said the same thing:

"Drug companies now finance most clinical research on prescription drugs, and there is mounting evidence that they often skew the research they sponsor to make their drugs look better and safer," said Dr Marcia Angell.

The report in the *Journal of Law, Medicine & Ethics* pings Gardasil's manufacturer Merck for skewing data:

"With regard to Gardasil, we noted that often in trials sponsored by the vaccine manufacturer, the assessment of the frequency of Adverse Drug Reactions was limited to those trial cohorts which comprised of participants who did not receive the full three doses of the HPV vaccine.

"The result of such population sample bias is a lesser sensitivity for detecting serious ADRs, as such events may be expected to occur less frequently if fewer doses of the vaccine are administered."

Then there's the hoopla and hype surrounding Gardasil. In a radio interview, the New Zealand Ministry of Health announced, "We'd just like to encourage young women that are listening to this to please go and have your Gardasil injection, because you will protect yourself for the rest of your life."

That comment, by the Ministry's Dr Api Talemaitoga to Newstalk ZB, seems a trifle over-optimistic, in the face of recent scientific evidence.

"Existing clinical trials show that antibodies against HPV-18 from Gardasil fall rapidly, with 35% of women having no measurable antibody titers at five years. This outcome suggests that rather than preventing future cases of cervical cancer, Gardasil may only be effective in postponing them," says the *Journal of Law, Medicine & Ethics* report.

If Gardasil only protects for a few years, and your daughter is immunised at 12, will she still be immune at 18 or 20, or is the Ministry of Health falsely promoting this vaccine as effective when it is not?

In fact, the latest studies have shown Gardasil appears to be making no impact on cervical cancer rates, beyond what the Pap smear programme has already achieved. The journal calls claims about Gardasil's effectiveness "highly misleading" when the figures are properly analysed.

Furthermore, because the HPV vaccine only works against a minority of the HPV viruses, there's every chance teenagers given the jab will falsely assume they are completely protected – "I've had the jab" – when in fact there's still a greater than 60% risk of infection from some of the other viruses.

"To date," notes the journal, "clinical trial evidence has not demonstrated that Gardasil can actually prevent cervical cancer (let alone cervical cancer deaths) because the follow-up period was too short (five years) while cervical cancer takes 20-40 years to develop…what Gardasil has been demonstrated to prevent are two out of 15 oncogenic HPV strains (HPV-16 and HPV-18)."

And we've already seen that immunity to the HPV-18 strain appears to be short-lived.

What is worrying is that the vaccine was given fast track approval for use on children despite a lack of evidence about safety or efficiency.

"Merck's HPV vaccine Gardasil failed (and continues to fail) to meet a single one of the four criteria required by the FDA for Fast Track approval," say Tomljenovic and Shaw. "Gardasil is demonstrably neither safer nor more effective than Pap screening combined with LEEP, nor can it improve the diagnosis of serious cervical cancer outcomes.

"In spite of this, Gardasil continues to be promoted as if it already had post-phase 4 confirmatory trial approval and proven efficacy against cervical cancer.

"Any federal agency responsible for assuring drug safety should not exclusively rely on data provided by the drug manufacturer, as unreliable research (ie, use of a reactive and potentially toxic placebo) cannot be used to reliably evaluate the safety of any drug."

The fact that one of the world's largest vaccine makers and drug companies is being accused of using substandard research should sound warning bells, especially given recent history of the major pharmaceutical companies. We are often forced to trust them with our lives. The question is, should we?

GlaxoSmithKline's woes are far from over. In September 2013 Chinese authorities announced they were considering "astronomical" fines against Glaxo over bribery allegations. The Ministry of Public Security in Beijing published some of the allegations from senior GSK staff in China on its website:

"In the past few years, GSK's UK headquarters gave us extremely high annual sales growth targets of as much as 25 per cent, far above the industry average of 7 to 8 per cent," said GSK China general manager Huang Hong. "With such unreasonable targets, if we do not resort to illegal measures, it's very difficult to achieve such high sales growth. Mark Reilly [former China CEO] changed the company's objective to sales being king."[71]

The overwhelming message from all of this is that you can no longer trust the assurances of the big pharmaceutical companies that their products are safe, that they have honestly and openly declared all relevant information from scientific testing, and that they are working for the

---

71 "Beijing weighing large fines against GlaxoSmithKline", South China Morning Post, 4 September 2013 http://www.scmp.com/news/china/article/1302852/beijing-weighing-large-fines-against-glaxosmithkline

health of you and your family. These companies are so big that even a $3 billion fine is water off a duck's back.

You may be wondering why state health agencies put so much faith in the giant pharmaceuticals. It appears they have little option. In New Zealand, senior politicians have privately indicated that Big Pharma is beyond their reach. The country, which has a socialised taxpayer-funded health system, relies on the giants to provide cheap medicines in return for giving them pretty much free rein. New Zealand is one of the few countries in the world where pharmaceutical companies can market direct to consumers in the mainstream media.

The pharmaceutical giants make small fortunes from national vaccination campaigns backed with the authority of the Ministry of Health. In fact, in Gardasil's case, the New Zealand Ministry of Health actually made large cash payments as an inducement to schools who allowed their students to be vaccinated.[72]

But it is not just national health policies that Big Pharma is influencing. They also play a major role in developing the United Nations international health and food policy, known as the Codex Alimentarius.

---

72 TGIF Edition, 27 March 2009, http://issuu.com/iwishart/docs/tgif27march09

CHAPTER 6

Food Control

*"Monsanto should not have to vouchsafe the safety of
biotech food. Our interest is in selling as much of it as
possible. Assuring its safety is the F.D.A.'s job"*
– Philip Angell, Monsanto 1998[73]

Codex Alimentarius. It's a Latin name that literally means "Book of
Food", but this is no recipe guide. Instead, the Codex was established by
the United Nations in 1963 to standardise world food rules.

The United Nations is destined to feature a great deal in this book,
but Codex is essentially about creating a global food economy, and it is
backed up by the sanctions power of the UN's World Trade Organisation.
In other words, if a country is non-compliant once ratifying Codex, it
can be punished internationally for breaking the rules.

In New Zealand, which is, as always, a cheerleader for United Nations
edicts, the Food Bill is being pushed through parliament to give legisla-
tive authority to the Codex.

"New Zealand attaches great importance to the work of Codex," says
the Ministry for Primary Industry in an overview.[74]

Among the powers it delegates, it creates Food Police who can:

Enter homes or businesses without notice if necessary

Enter a place without a search warrant and use any force that is reasonable

73 "Playing God in the Garden" New York Times Magazine, October 25, 1998
74 http://www.foodsafety.govt.nz/policy-law/codex/nz/index.htm

Even the police don't have power to enter and search homes without a warrant, unless they have reason to believe a life is in danger or drugs are on the premises. But getting busted for baking muffins without a permit?

If you grow food in your garden and trade some with your neighbours, you could be deemed to be a food producer under the new law:

"This food sector covers food businesses that are horticultural producers (farmers or growers) of fruit, vegetables, herbs, spices, nuts, cereal grains, seeds, fungi, grasses, or any components extracted or gathered from horticultural produce."

More significantly under that particular clause, the Codex establishes legal controls over seeds and horticultural production. For millennia, humans have had a natural right to grow and produce food. Now, the United Nations is proposing – and getting – global regulations introduced to control who is allowed to produce food.

What has long been a basic human right for millennia, to grow and market your own produce, suddenly becomes a "privilege" conferred on you by the State. Are there hundreds of thousands of New Zealanders dropping dead from food poisoning at a sausage sizzle every year? No, serious food poisoning is so rare in New Zealand it makes the national headlines once every three decades.

It's not as if governments around the world, or the United Nations, asked the public for the right to usurp this authority to grow your own. They just assumed the power and did it. That, of course, is the heart of a totalitarian state of mind, feel the power and do it anyway.

The money, the payoff, comes from the implementation of such a system globally because it means the multinationals can carve up the third world markets where most of the growth will come from this century. The real money in controlling the world won't be made in America or New Zealand or the UK – it will be made in Africa, Asia, South America and the Middle East. But those markets can't be fully exploited until a world governance structure is put in place that protects the exploiters from the whims of sovereign countries. Once a country ratifies Codex – like any UN treaty – it is in for life, and subject to crushing trade sanctions if it doesn't obey the rules.

This, of course, plays right into the hands of global multinationals like Dow Agrosciences and Monsanto. Both of these companies are perhaps best known for producing toxic chemicals – herbicides and pesticides. But let's take a look at how Monsanto has diversified. It began as the company that invented the artificial sweetener saccharin. It invented Astro-turf and also was the first to mass-produce LEDs (light-emitting diodes). Moving into agricultural chemicals it manufactured the components of Agent Orange – used to devastating effect in the Vietnam War but with equally devastating birth deformities among the families of all those exposed to it.

Monsanto's board of directors includes the head of McDonalds USA, and a top executive with Procter & Gamble.

In this book, you will meet bureaucratic totalitarians, political totalitarians, spiritual totalitarians and corporate totalitarians. Monsanto and the other big global brands are examples of the corporate variety.

Despite using the phrase "Sustainable Agriculture" on its website, the company has an exceedingly bad reputation. It manufactured nearly 100% of the toxic PCBs now polluting the USA. Its manufacturing operations in the UK created a toxic waste dump as well. It plays a huge role however in guiding global food, agriculture and chemical rules.

Although Monsanto's corporate website only contains a glancing reference to Codex in two documents, the Codex website contains 379 documents featuring Monsanto. The multinational works through a range of front organisations, like the International Life Sciences Institute, or the Alliance to Feed The Future. These are all "soft faces" for Monsanto PR executives to manipulate the system.

The Alliance, for example, is controlled by an outfit called the International Food Information Council. When you go to the IFIC website, if you know where to dig (it is not easy), you will eventually find it is funded by a group of global multinationals including:[75]

- Monsanto
- DuPont
- Abbott Nutrition
- Bayer Crop Science
- Dow AgroSciences
- Coca Cola
- Dr Pepper
- General Mills
- Hershey

---

75 http://www.foodinsight.org/linkclick.aspx?fileticket=09a8FrY%2bp7M%3d&tabid=91

- Heinz
- Kelloggs
- Kraft
- Mars
- McDonalds
- Nestle
- Pepsi
- Red Bull
- Unilever

The International Food Information Council sees its role as "educating" the public and the media about nutrition and food safety. "We also serve as a news media resource. We provide science-based information to the media and refer journalists to our 350 independent, credentialed experts on a variety of nutrition and food safety topics."[76]

It's a very slick PR campaign, complete with soundbites delivered on the issues of the day to media everywhere.

Its "partners" include a veritable who's who of global food and health organisations. The outlets endorse the IFIC, in fact the American Academy of Physicians Assistants even endorsed a puff piece brochure advising consumers that the artificial sweetener aspartame is perfectly safe:

"Aspartame has been studied extensively and has been found to be safe by experts and researchers. Government agencies worldwide, including the U.S. Food and Drug Administration (FDA), have also reviewed the science and found aspartame to be safe for human consumption."[77]

Aspartame was once one of Monsanto's mainstay products, you may know it better by its trade names NutraSweet and Equal, among others. It is surprising that the statement above about it being "studied extensively" made it through peer review. In 2012 the *American Journal of Clinical Nutrition* published a study linking aspartame use in humans to a higher risk of cancer, and it was critical at how "sparse" previous human studies had been and how poorly designed those studies were, with "only short term follow-up". In simple terms, if you hadn't developed cancer within a short time of drinking diet soda, that proved it was "safe" for the purposes of the food industry.

76 http://www.foodinsight.org/about-ific-and-food-safety.aspx
77 http://www.foodinsight.org/Content/3848/FINAL_Aspartame%20Brochure_Web%20Version_11-2011.pdf

This new study followed diet soda drinkers for up to 20 years and found the risk of cancer more than doubled for some tumours.[78] Another study of 59,000 Danish pregnant women has found consumption of diet soft drinks strongly increased the risk of having an early premature baby.[79]

It turns out that virtually all the studies that have found aspartame to be safe were funded by or linked to the interests of aspartame manufacturers – exactly the same shonky science we saw with Big Pharmaceuticals. As Professor Ralph G. Walton, M.D., Professor of Clinical Psychiatry, at the USA's Northeastern Ohio Universities College of Medicine, warned back in 2003:

"The diet food industry and the F.D.A. are fond of saying that aspartame is 'the most studied product in history' with an outstanding safety record. In fact however virtually all of the studies in the medical literature attesting to its safety were funded by the industry[80], whereas independently funded studies, now numbering close to 100, identify one or more problems."[81]

Aspartame, incidentally, rapidly breaks down at body temperature to three chemicals, all of which are toxic to human health: methanol (the toxic poison in methylated spirits) 10%, Aspartic acid, (around 40%); Phenylalanine, (around 50%). It does a number of things, including affecting the central nervous system by acting as an 'excitotoxin'.

The aspartame controversy deserves a book of its own, but the preceding few paragraphs should be sufficient to show you, first hand, how big chemical companies are using their links to official health agencies to mislead you about the safety of their food additives. They simply keep lying to you with all the calm assurance that authority figures routinely muster. They lie to you directly, and they lie through the news media. The fact is they are doing it with the assistance of the totalitariacrats, because they're part of that globalisation movement.

Sometimes it isn't additives but residues. For two decades, Monsanto, the chemical giant that manufactures Roundup (glyphosate) has been developing genetically modified food crops, and getting them authorised

78 "Soft drinks, aspartame, and the risk of cancer and cardiovascular disease", Am J Clin Nutr December 2012 vol. 96 no. 6 1249-1251
79 "Intake of artificially sweetened soft drinks and risk of preterm delivery: a prospective cohort study in 59,334 Danish pregnant women", Halldorsson et al, 2010, doi: 10.3945/ajcn.2009.28968 Am J Clin Nutr September 2010 vol. 92 no. 3 626-633
80 http://www.dorway.com/peerrev.html
81 http://www.thebriefingroom.com/archives/2007/11/aspartame_sweet.html

for use in the food chain. The point of these crops was not modification to make them more nutritious, but to make them more resistant
to herbicides – specifically Monsanto's own herbicide Roundup – or in
some cases the crops were engineered to produce their own pesticide. For
farmers, that means they can dump greater quantities of the weed-killer
Roundup on the crops secure in the knowledge it will get rid of the weeds
but not kill the crop.

Monsanto wins two ways: firstly, it sells vastly more Roundup. Secondly,
it owns the rights to the genetically-modified crop seeds, so that farmers
are forced to buy seed from Monsanto each year.

The chemical giant aggressively pushed for legal recognition of its intellectual property rights, getting courts in the US and Canada to agree
that where Monsanto's GM seed spread into neighbouring farms growing
natural crops, that neighbouring farmers be forced to hand their crops
over to Monsanto for breach of the patented technology.

Monsanto has also developed so-called "terminator seed" technology.
This is food crop seed that is sterile, meaning that it will grow the current
year's crop, but all seeds from that harvest are barren. Given how easily
the GM seed varieties have been shown to spread in nature, the dangers
of a terminator gene escaping into natural food crops could cause massive
food shortages and starvation.

Recognising it didn't have a watertight argument to debunk those
fears, Monsanto has promised not to commercialise its terminator seeds.
Nonetheless, it hangs unspoken.

For consumers, however, GM crops mean greater exposure to the herbicide Roundup in your food. A study of 182 people in 18 European cities
found Roundup in the urine of 44% of those sampled. The significant fact
was that these were people who had no daily contact with the herbicide.[82]

Of more concern, for decades Monsanto has claimed Roundup is safe
and wheeled out scientific studies supposedly proving that. Given the
recent revelations about scientific fraud by big pharmaceutical companies,
that may be reason for caution. Fears rise considerably when some newer
studies are added to the mix. A 2012 study on rats fed Roundup-resistant
GM corn found death rates doubled or tripled, thanks to cancer and
kidney damage.

"A Roundup-tolerant maize and Roundup provoked chronic hormone

82 http://blogs.wsj.com/brussels/2013/06/13/study-youre-in-trouble-roundup/

and sex dependent pathologies. Female mortality was 2–3 times increased mostly due to large mammary tumors and disabled pituitary. Males had liver congestions, necrosis, severe kidney nephropathies and large palpable tumors. This may be due to an endocrine disruption linked to Roundup and a new metabolism due to the transgene. GMOs and formulated pesticides must be evaluated by long term studies to measure toxic effects."[83]

That study was quickly attacked by official Food Safety agencies in Europe, Australia and New Zealand[84], who claimed the cancers and other rat defects could be explained by natural variation. Given that glyphosate is now the world's most widely-used herbicide, and manufactured by a range of massive multinationals, there is a huge amount riding on this research, and the industry is certainly not standing by to see its golden goose killed off.

The Food Safety agencies are creatures of the United Nations Codex, and Codex and its national agencies – as we have established – have Monsanto and the other massive multinationals as key technical advisors.

Case in point, one of Monsanto's top lawyers, Michael Taylor, was appointed at senior advisor to the Food and Drug Administration in the US, making him, as the *Huffington Post* reported, "now America's food safety czar. What have we done?"[85]

The *Huff* linked to leaked documents disclosing how Monsanto and the regulators had covered up suspected toxicities in the GM food ranges.[86]

Monsanto and the pharmaceutical giants are owned, for the most part, by exactly the same massive Wall Street investment funds and merchant banks. It is as if western economic power is concentrated in Big Pharma and Big Chemical, and companies like Monsanto, Glaxo or Johnson & Johnson are little more than brand-differentiated sock puppets of the same shareholders.[87]

Curiously, many of the scientists who attacked the Seralini research appear to have undisclosed conflicts of interest, linking them to the chemical giant

83 "Long term toxicity of a Roundup herbicide and a Roundup-tolerant genetically modified maize", Seralini et al, Food and Chemical Toxicology, Volume 50, Issue 11, November 2012, Pages 4221–4231
84 http://www.foodstandards.govt.nz/consumer/gmfood/seralini/pages/default.aspx
85 "You're Appointing Who? Please Obama, Say It's Not So!" by Jeffrey Smith, Huffington Post, 23 July 2009, http://www.huffingtonpost.com/jeffrey-smith/youre-appointing-who-plea_b_243810.html
86 http://biointegrity.org/index.htm
87 Compare shareholdings for Monsanto, Johnson & Johnson and Glaxo here: http://finance.yahoo.com/q/mh?s=MON+Major+Holders and here: http://finance.yahoo.com/q/mh?s=JNJ+Major+Holders and here: http://finance.yahoo.com/q/mh?s=GSK+Major+Holders

or its allies. An example was Paul Christou, who headed a rival study in the journal *Transgenic Research* claiming to have debunked Seralini's links with GM Roundup-ready corn to cancers and kidney damage.

"Christou's article attacking Séralini's study carried no such disclosure. Yet his conflicts of interest are serious and extensive", writes GM activist and *Spinwatch* editor Claire Robinson.[88]

"He worked for the GM seed company Agracetus for 12 years until 1994. Agracetus's biotechnology interests were bought up by Monsanto in 1996. Christou is an inventor on a number of patents on GM crop technology, for most of which Monsanto owns the property rights. It is not known whether Christou earns money from these patents but even if he does not, it is normal practice to declare inventor status on patents as a competing interest in scientific articles.

"From 1994 to 2004, Christou worked at the John Innes Centre in the UK, which is also heavily invested in GM crop technology."

As explained, there is huge investment riding on this issue of controlling and patenting food production.

Former US Agriculture Secretary Dan Glickman served in the position from 1995 until 2001, at the height of the GM debate:

"What I saw generically on the pro-biotech side was the attitude that the technology was good and that it was almost immoral to say that it wasn't good because it was going to solve the problems of the human race and feed the hungry and clothe the naked. And there was a lot of money that had been invested in this, and if you're against it, you're Luddites, you're stupid. There was rhetoric like that even here in this department. You felt like you were almost an alien, disloyal, by trying to present an open-minded view on some of the issues being raised. So I pretty much spouted the rhetoric that everybody else around here spouted; it was written into my speeches".[89]

And you thought food regulation was objective, scientific and done on its merits?

You may wonder why the mainstream media don't do more to expose the conflicts of interest. One of the reasons – in the UK, Australia and

---

88 "Tumorous rats, GM contamination, and hidden conflicts of interest", by Claire Robinson, Spinwatch, 30 May 2013. Full footnotes for her claims can be found in the linked story: http://www.spinwatch.org/index.php/issues/more/item/5495-tumorous-rats-gm-contamination-and-hidden-conflicts-of-interest

89 "Outgoing Secretary Says Agency's Top Issue is Genetically Modified Food", by Bill Lambrecht, St. Louis Post-Dispatch, 25 January 2001

New Zealand at least – is the existence of a propaganda unit known as the "Science Media Centre" or SMC. The SMC's New Zealand operation, as an example of the type, is corrupt.[90] It was established to "brief" the news media on scientific developments, but on controversial issues runs a clear establishment line and blocks its critics from asking inconvenient questions at news conferences.

Journalists are spoon-fed quotes from the SMC's chosen "experts", and the story is accordingly spun. Within 24 hours of the Seralini paper's publication, the SMCs of the UK, Australia and New Zealand were running the same interference, including providing quotes from scientists who had what critics called undisclosed conflicts of interest.

The Science Media Centre, associated with the Royal Society, is effectively a front organisation not dissimilar to the International Food Information Council and its propaganda wing.

Regardless of the attacks on the Seralini study, another from Thailand has found glyphosate fuels the proliferation of breast cancer cells in women, meaning it could be causing a rise in breast cancer because it is present in so many foods.[91]

Some of the GM crops contain Bt toxins, a bacterial byproduct that kills some insect pests, but new research has found those toxins also kill red blood cells. The same study says the toxins have been found in the blood of women, both pregnant and not, and may contribute towards anemia.[92]

Perhaps the biggest blow to the credibility of Roundup, however, came in a recent study from MIT in Boston.[93]

"Glyphosate, the active ingredient in Roundup, is the most popular herbicide used worldwide. The industry asserts it is minimally toxic to humans, but here we argue otherwise," says the MIT team. "Residues are found in the main foods of the Western diet, comprised primarily of sugar, corn, soy and wheat."

---

90 For the factual context behind this opinion, see http://briefingroom.typepad.com/the_briefing_room/2009/11/science-media-centre-cant-take-the-heat-on-climate-change.html
91 "Glyphosate induces human breast cancer cells growth via estrogen receptors", Thongprakaisang et al, Food Chem Toxicol. 2013 Sep;59:129-36. doi: 10.1016/j.fct.2013.05.057. Epub 2013 Jun 10, http://www.ncbi.nlm.nih.gov/pubmed/23756170
92 "Hematotoxicity of Bacillus thuringiensis as Spore-crystal Strains Cry1Aa, Cry1Ab, Cry1Ac or Cry2Aa in Swiss Albino Mice", Mezzomo et al, Journal of Hematology & Thromboembolic Diseases, 2013, 1:1 http://dx.doi.org/10.4172/jhtd.1000104
http://gmoevidence.com/wp-content/uploads/2013/05/JHTD-1-104.pdf
93 "Glyphosate's Suppression of Cytochrome P450 Enzymes and Amino Acid Biosynthesis by the Gut Microbiome: Pathways to Modern Diseases", Anthony Samsel and Stephanie Seneff, Entropy 2013, 15(4), 1416-1463; doi:10.3390/e15041416, http://www.mdpi.com/1099-4300/15/4/1416/pdf

Previous studies that declared glyphosate "safe" studiously ignored glyphosate's inhibition of a certain enzyme important in mammals. This study decided to look a little closer. They found the enzyme plays a huge role in protecting against a range of disorders. With Roundup circulating in the blood the enzyme is weakened:

"Consequences are most of the diseases and conditions associated with a Western diet, which include gastrointestinal disorders, obesity, diabetes, heart disease, depression, autism, infertility, cancer and Alzheimer's disease. We explain the documented effects of glyphosate and its ability to induce disease, and we show that glyphosate is the 'textbook example' of …disruption…by environmental toxins."

In plain English: food sprayed with Roundup (which is most commercially produced plant food) appears to be poisonous. The cynical amongst you have probably already realised that if the same shareholders who own the companies poisoning your food also own the companies making expensive medicines, then they are managing to clip the ticket at both ends of the deal.

"Contrary to the current widely-held misconception that glyphosate is relatively harmless to humans, the available evidence shows that glyphosate may rather be the most important factor in the development of multiple chronic diseases and conditions that have become prevalent in Westernized societies," the MIT study notes.

To get an idea of just how big this timebomb could be, consider these US Department of Agriculture figures. In 2000, just 7% of the national maize crop was genetically-modified to be Roundup-resistant. By 2012, 73% of corn was Roundup-ready.[94]

Another study found farmers were now able to apply between 200 and 500% more Roundup on their crops than they had previously done – five times more of the chemical.

No wonder Monsanto and the other Roundup makers like Dow and Syngenta are smiling.

Some of you right now might be asking, hang on, if the Food Police are being given power to raid your kitchen and squeeze your muffins, surely they have the power to enforce the safety of genetically-modified food? The short answer is no, they don't.[95]

---

94 http://www.ers.usda.gov/datafiles/Adoption_of_Genetically_Engineered_Crops_in_the_US/alltables.xls
95 In fact, the US Government snuck in a provision in March 2013 under the cover of a budget bill,

As it currently stands under the UN food Codex that the multinationals are operating under, responsibility for testing the safety of GM food ingredients is left in the hands of the multinationals and their own research labs. There is to be no independent oversight by the Food Police.

"The stance taken by Monsanto, Dow and the other peddlers of both chemicals and genetically engineered seeds is that GMO food is 'identical to non-GMO products'. They claim that genetic engineering is no different than plant hybridization, which has been practiced for centuries. It is the reason they gave, and the EPA accepted, for not having to submit GMO food to rigorous testing to obtain EPA approval. It's up to the companies that manufacture GMOs to research and determine the safety of their products," says GMO expert and commentator Dr Nancy Swanson.[96] She's not a medical doctor, but a retired university physics professor and US Navy scientist.[97]

The idea that genetic modification is the same as cross-breeding and 'hybridization' is something she finds laughable:

"Not only are the bacteria genes [inserted into plants] themselves potentially toxic, but the plants can be sprayed directly with herbicides, the herbicide-resistant plants absorb the poisons and we eat them. It's difficult to understand how this can be considered 'essentially' the same as plant hybridization."

And all of this, without any genuine independent testing of the organisms.

In fact, the Monsanto/Bayer/Dow/DuPont sponsored International Food Information Council boasts about its ability to control its own safety testing under what is known as GRAS, or Generally Regarded As Safe regime.

In a powerpoint presentation, delegates are told that using the GRAS regime to get food additives approved in the US is much faster and simpler than going through the formal petition process with the Food and Drug Administration.

Under a question, "Who makes Safety Determination?", delegates are told:

---

that made Monsanto and other big chemical companies legally immune in the event their genetic modification injures the public. It's been dubbed the "Monsanto Protection Act", and specifically is Clause 735 of the bill HR 933 passed on 22 March 2013. As Salon magazine wrote: "The provision protects genetically modified seeds from litigation in the face of health risks...President Barack Obama signed the spending bill, including the provision, into law on Tuesday." For a link to the actual legislation, and the background, see Salon: http://www.salon.com/2013/03/27/how_the_monsanto_protection_act_snuck_into_law/
96 http://www.examiner.com/article/washington-state-residents-likely-to-vote-on-gmo-food-labels-1
97 http://www.examiner.com/gmo-in-seattle/nancy-swanson

"Submitter of Notification; Uses Experts; FDA Issues No Objection Letter."[98]

The process takes "days" or "months", compared to "years" under the formal petition process. Of course, if you control the experts, the whole process is plain sailing.

And that's exactly what an independent review has just found. In the *Journal of the American Medical Association* "Internal Medicine" issue in August 2013, researchers found the Food and Drug Administration doesn't even know what many of the food additive chemicals actually are! The GRAS regime dates back to 1958, and it has been used to approve around 43% of the 10,000 chemicals known to have been OK'd for use in food.

What's worse, of those 10,000 chemicals, a thousand of them have never been disclosed to the FDA at all. The FDA is completely in the dark about precisely what those chemicals might be.

The JAMA team analysed 451 GRAS applications made between 1997 and 2012 and published on the FDA's website. What they found shows you just how total the control of the chemical companies is over official food safety – not one of the 451 Generally Regarded As Safe approvals was made by an independent expert. Not one:

"About 22 percent were made by an employee of the food additive manufacturer and about 13 percent by employees of consulting firms hired by manufacturers," reported Reuters.[99]

"Another 64 percent were submitted by food safety expert panels whose members were picked by either manufacturers or consulting firms to evaluate the additive. None of the panels, which included an average of four people, were selected by third parties, the study found.

"Neltner's study also found that 10 experts served on 27 or more panels. One of them was a member of 128 panels, about 44 percent of the total."

Another study in the journal *Reproductive Toxicology* surveyed the FDA food additive database and found that there was insufficient toxicology data on 80% of the additives to illustrate what the safe human consumption limits of the chemicals actually are.[100]

---

98 http://www.foodadditives.org/pdf/Safety%20and%20Suitability%20of%20Food%20Additives.pdf
99 "Industry has "undue influence" over U.S. food additives: study", Reuters, 7 August 2013 http://www.reuters.com/article/2013/08/07/us-food-additives-bias-idUSBRE9760Z820130807
100 "Data gaps in toxicity testing of chemicals allowed in food in the United States", Thomas G. Neltner et al, Reproductive Toxicology, Volume 42, December 2013, Pages 85–94 http://www.sciencedirect.com/science/article/pii/S0890623813003298

In fact, of the 10,000 or so known food chemicals, no controlled safety tests have been done on a large majority, says the new report:

"Almost two-thirds of chemical additives appear to have been declared safe for use in food without the benefit of being fed to an animal in a controlled toxicology study."

What are the implications for the health of you and your family? In short, the chemicals added to processed foods may well be toxic to adults and children, because no real safety data has ever been provided: "given the substantial lack of toxicology data for chemical additives, extrapolation of the limited information to so many chemicals is disconcerting and may be insufficient to ensure safety. Therefore, it may represent a public health problem.

"With almost two-thirds of chemical additives lacking feeding toxicology and 78.4% of additives directly added to food lacking data to estimate a safe level of exposure and 93% lacking reproductive or development toxicity testing, it is problematic to assert that we know with reasonable certainty that all chemical additives are safe.

"Although FDA is aware of the problem, it lacks the authority and resources to fill the information gaps. Furthermore, once a chemical is approved, manufacturers have no incentive to add additional toxicology information because FDA neither has a reassessment program in place nor has authority to require additional testing."

In response, the Grocery Manufacturers Association in America defended the GRAS system, telling the media it was a "thorough and comprehensive process". Believe that, and you will believe pigs can fly. Given Monsanto's involvement in genetic engineering of pigs, however, that does remain a possibility.

You can read a list of the Food Additive numbers for some of the additives on Wikipedia if you are curious.

These global food industry bodies representing the chemical companies and processed food manufacturers all insist they are "compliant" with the UN's Codex Alimentarius. The shocking thing is, they're not lying, they are compliant with UN edicts because to a large extent they control Codex. They are drafting food laws across the world[101] that heavily favour globalised

---

101 Among them, the global corporations are united in opposing food product labelling laws. Calfifornia's Prop. 37 law seeking mandatory disclosure of genetically modified content on food labels was opposed by Monsanto, DuPont, Hersheys, Mars, Kraft, PepsiCo and Syngenta, to name a few, who donated tens of millions of dollars to fund advertising campaigns against GM labelling. Their campaign succeeded, voters opted not to support labelling. Similar political/corporate

food processors, but which will heavily punish any homeowners, farmers or small businesses trying to compete by offering more natural food.

It is no coincidence that New Zealand's Codex-compliant Food Bill specifically excluded genetically modified food from the items the Food Police have jurisdiction over.[102] It is also no coincidence that these multinational companies make some of the biggest profits on the planet.

To add insult to injury, five years after making a $5 million donation to the World Food Prize Foundation, Monsanto this year won the event through one of its top GMO scientists, leaving critics speechless:

"Winning this prize will encourage the wider use of genetically engineered crops and be a huge obstacle to those fighting to investigate the long-term effects of its frankenseeds – which is exactly what Monsanto wants," said activist Oliver Moldenhauer. "In 2008, Monsanto made a $5 million pledge to the World Food Prize Foundation, part of its plan to buy the credibility it can't legitimately earn. By handing its benefactor this award, the Foundation risks undermining the credibility of the most respected prize in agriculture."[103]

One of the scientists who shared the prize worked for Syngenta, whose pesticides have been accused in Europe of contributing to the demise of bee colonies.

Do you still trust the United Nations and your national food regulatory agency, or are you beginning to suspect you are being conned? By the time you reach the end of this book, the answer will have been writ large in front of you.

The way Codex treats natural health supplements like vitamins and minerals stands in sharp contrast to the slack standards on other food chemicals. While Monsanto, Dow, DuPont and others can sneak chemicals into food with no controlled safety tests whatsoever, they influenced Codex to require full scientific testing of vitamin and mineral supplements, as the Codex guidelines make clear:

---

opposition to labelling is found in New Zealand and other Codex-dominated countries. http://www.huffingtonpost.com/2012/11/02/prop-37-donors-revealed-f_n_2065789.html

102 After public pressure, the Government agreed in 2013 to include a clause in the Bill permitting the New Zealand government to make its own laws on GM food. This was not included in the original international treaty guiding the process and has to be managed under the "exceptional circumstances" clause of the international treaty. Even so, Food Minister Nikki Kaye says "The Food Bill supports the existing robust scientific pre-approval process required prior to GM foods being allowed for sale". As you have seen, the process is far from "robust". http://www.scoop.co.nz/stories/PA1306/S00184/food-bill-changes-better-balance-legislation.htm

103 http://blogs.vancouversun.com/2013/10/15/consumer-group-outraged-at-monsanto-winning-nobel-prize-of-agriculture/

"Vitamin and mineral food supplements should contain vitamins/provitamins and minerals whose nutritional value for human beings has been proven by scientific data and whose status as vitamins and minerals is recognised by FAO [UN Food and Agriculture Organisation] and WHO [UN World Health Organisation]."[104]

At the same time, the big players, backed by the US and Food Standards Australia New Zealand (FSANZ), tried to encourage Codex not to cut the food additive aluminium from the approved list. You didn't know you were eating the toxic metal that's been linked to Alzheimers and other nasties? Don't worry, few people do know. One who does is US attorney Scott Tips, the general counsel for the National Health Federation in the US. The NHF is the only accredited consumer watchdog group permitted to attend Codex meetings, and Tips says the battle over aluminium is typical of the struggle.[105]

"Aluminum is a known neurotoxin, easily crossing the blood-brain barrier, and it interferes with ATP enzymes, which carry out the important function of energy transfer among brain cells. Aluminum worsens the effects of other toxins, such as pesticides, herbicides, mercury, cadmium, fluoride, lead, and glutamate. It also detaches highly oxidizing iron in the bloodstream from its protective carrier transferrin. This greatly increases the toxicity of iron and is at least one of the mechanisms by which aluminum is toxic to the brain. Warnings about the toxic effects of aluminum could, and do, fill volumes.

"Aluminum ammonium sulfate, aluminum silicate, calcium aluminum silicate, sodium aluminum phosphates, and sodium aluminosilicate are the food additives that Codex was reviewing this session. They can be found in practically as many foods as you can imagine: vegetables, soybean paste, crackers, pastas and noodles, bagels, English muffins, pita bread, bread and baking mixes, chewing gum, milk and cream powder, processed cheeses, flours, batters for fish and poultry, dairy-based drinks such as eggnog, beverage whiteners, dried-whey products, salt, seasonings and condiments, soup and broth mixes, and sauces. And do not think that you can always look at labels and see them disclosed there because often the aluminum compound is hidden within a particular product identity."

The Aluminum Association of American discloses on its website that

---

it was not happy when NHF managed to get the tolerable weekly limit of aluminium intake for humans lowered from 7mg a week to 1mg, per kilogram of body weight.

"In June of 2006, the Codex scientific group, JECFA, lowered the PTWI for aluminum from all sources from 7mg/kgbw to 1mg/kgbw. The aluminum industry has actively engaged the JECFA to re-assess the PTWI, based on new research completed by the industry to comply with emerging global regulations.

"A coalition of associations representing both aluminum producers and food additives manufacturers has completed research addressing data gaps that led to the downward revision of the PTWI and will present that research to the Codex group," said the Aluminum Association to its members in 2011.

Their campaign was successful. The tolerable limit of aluminium was doubled to 2mg per kilo of body weight per week, with the approval of the World Health Organisation.[106]

The European Food Safety Authority did its own research in 2008 and revealed children are at highest risk of exposure to aluminium, because of their small body weight but big appetites, and because the metal is present in most processed foods and infant milk formula, where levels reached 0.9 mg/kg of body weight per week for dairy powder, and 1.1mg/kg for soy formula.[107]

Some big brand infant formulas, it noted, contained four times that amount:

"The Panel noted that in some individual brands of formulae (both milk-based and soya-based) the aluminium concentration was around 4 times higher that the mean concentrations estimated above, leading to a 4 times higher potential exposure in brand-loyal infants."

In keeping with the "breast is best" message, aluminium levels were a staggering 63 times lower in breast milk:

"Potential exposure in breast-fed infants was estimated to be less than 0.07 mg/kg bw/week."

The European Food Safety Authority reports that "Aluminium can

---

106 "WHO Committee Ups Recommended Intake Limit for Aluminum in Diet", Aluminum Association, http://www.aluminum.org/AM/Template.cfm?Section=Weekly_Briefing&Template=/CM/HTMLDisplay.cfm&ContentID=31824
107 "Safety of aluminium from dietary intake: Scientific Opinion of the Panel on Food Additives, Flavourings, Processing Aids and Food Contact Materials (AFC)European Food Safety Authority, 2007 The EFSA Journal (2008) 754, 1-34, http://www.efsa.europa.eu/en/efsajournal/doc/754.pdf

enter the brain and reach the placenta and fetus," and that it remains in the body for "a very long time in various organs and tissues".

What's interesting is that the EFSA noted that there were next to no studies on the safety of aluminium in human diet, so it opted for an upper limit of 1mg/kgbw per week. It reached that decision in a 122 page report analysing the science. Despite that, the World Health Organisation after lobbying from the aluminium and food additives industry doubled the permissible levels.

It's worth taking a look at what quantities of this toxic metal the United Nations approved for public consumption. Up until 2006, based on the previous 7mg limit, an average 85kg male was absorbing 31 grams of aluminium into his body each year through food, deodorant and a range of other avenues and products. That's around 1.9kg of aluminium you've ingested by the age of 70. And they wonder why dementia sets in.

Under the new limit, someone born today will absorb more than half a kilogram of this metal over their life time.

But here's something else you didn't know: those figures are the "recommended" intake levels. Depending on how much processed food you eat, how much tinfoil you use in your food preparation and the like, your actual exposure each week, according to the EFSA can actually be this high:

"The results ranged from 18.6 to 156.2 mg/kg bw/week at the mean and from 35.3 to 286.8 mg/kg bw/week at the 95th percentile."[108]

As the EFSA forlornly noted, the actual intakes of dietary aluminium "exceed" by an order of magnitude the tolerable limits.

Putting that in plain English, a person at the upper range could absorb *1.27kg of aluminium a year*, or 89kg over a lifetime. That's equivalent to casting a statue of you in aluminium, or auditioning as the Tin Man in the Wizard of Oz. If all that aluminium from just one year was concentrated in one lump in your body you would set off the metal detectors at airports. If it was magnetic, you would be a serious navigational hazard on a plane.

The studies keep mounting up, with one recent animal testing experiment concluding that exposure to aluminium, "particularly during pregnancy and lactation period, can affect the in utero developing fetus and

---

108 "Dietary exposure to aluminium-containing food additives", Question number: EFSA-Q-2013-00312
Issued: 13 March 2013, http://www.efsa.europa.eu/en/supporting/pub/411e.htm

postnatal…raising the concerns that during a critical perinatal period of brain development, Al exposure has potential and long lasting neurotoxic hazards."[109]

Another new study on human brains reinforces the disturbing message: "Once biologically available aluminum bypasses gastrointestinal and blood–brain barriers, this environmentally-abundant neurotoxin has an exceedingly high affinity for the large pyramidal neurons of the human brain hippocampus. This same anatomical region of the brain is also targeted by the earliest evidence of Alzheimer's disease (AD) neuropathology."[110]

The study reports that "human brain endothelial cells were found to have an extremely high affinity for aluminium", meaning that the food you are feeding yourself or your children, approved by the United Nations Codex and enforced by your local food laws, may be giving you a very nasty death sentence in the form of eventual dementia.

Good to know the chemical companies have been adding it to our food. Not so good to know that FSANZ downunder and FDA in the USA have been complicit in this. It probably won't come as a surprise to learn that "to this date, aluminum has never been tested for safety by the FDA," says Tips.

But it's not just aluminium they've been dosing the public with.

109 "Neurobehavioral toxic effects of perinatal oral exposure to aluminum on the developmental motor reflexes, learning, memory and brain neurotransmitters of mice offspring," Gasem M. Abu-Taweela et al, Pharmacology Biochemistry and Behavior, Volume 101, Issue 1, March 2012, Pages 49–56
110 "Selective accumulation of aluminum in cerebral arteries in Alzheimer's disease (AD)" Bhattacharjeea et al, Journal of Inorganic Biochemistry, Volume 126, September 2013, Pages 35–37, http://www.sciencedirect.com/science/article/pii/S0162013413001207

CHAPTER 7

Fluoride: It Does Get In!

"Fluoridation...is the greatest fraud that has ever been perpetrated and it has been perpetrated on more people than any other fraud has."
– Dr. Professor Albert Schatz, microbiologist and co-discoverer of Streptomycin

Millions of people have fluoride added to their drinking water by the state. It is done in the name of of preventing tooth decay, but the Nazis and others knew that fluoride also dumbs populations down and makes them more docile.

Studies on fluoride, incidentally, have found a similar targeting of the hippocampus area of the brain as that seen with alumnium, leading to growing scientific speculation that fluoride at drinking water doses may have a long term effect on intelligence and mental function.[111] A recent rat study backs that up.[112]

"We found that NaF (sodium fluoride) treatment impaired learning and memory in these rats. Furthermore, NaF caused neuronal degeneration, decreased brain glucose utilization, decreased the protein expression of glucose transporter 1 and glial fibrillary acidic protein, and increased levels

---

111 http://interesjournal.com/JMMS/Pdf/2012/April/Wilson%20et%20al.pdf
112 "Low Glucose Utilization and Neurodegenerative Changes Caused by Sodium Fluoride Exposure in Rat's Developmental Brain", Jiang et al, NeuroMolecular Medicine, August 2013, DOI: 10.1007/s12017-013-8260-z

of brain-derived neurotrophic factor in the rat brains. The developmental neurotoxicity of fluoride may be closely associated with low glucose utilization and neurodegenerative changes."

Fluoride can be safely classed as rat poison based on a barrage of similar studies, yet many of us drink it every day, it's in our food as a result of water fluoridation, and some people even vote to add it to their drinking water![113]

If you are a parent and you support fluoride in drinking water because government health agencies tell you it's a good thing, spare a thought for a comparison study in India between areas of low fluoridation and high fluoridation, and the impact on the IQ of children:

"The average IQ level of schoolchildren (N = 50) from the low F villages was 97.17, which is significantly higher (p≤0.001) than 92.53 of schoolchildren (N = 34) from the high F villages."[114]

Another peer reviewed study in 2013 compared 12 year olds in areas with less than 1.5 parts per million of fluoride in the water, compared to areas above that level. They too found a drop in IQ: "Reduction in intelligence was observed with an increased water fluoride level."[115]

An Iranian study looked at children in low, medium and high fluoride areas, and found an IQ drop of nearly 10 points for children getting plenty of fluoride, which is absolutely massive. Licking lead paint on a child's cot resulted in a three point drop in IQ, by way of comparison.[116]

"The mean IQ scores decreased from 97.77±18.91 for the normal fluoride group to 89.03±12.99 for the medium fluoride group and to 88.58±16.01 for the high fluoride group (P=0.001)."[117]

---

113 "Hamilton, Whakatane and Hastings want water fluoridated", TVNZ, 12 October 2013 http://tvnz.co.nz/national-news/hamilton-whakatane-and-hastings-want-water-fluoridated-5645394
114 "Fluoride contamination of groundwater and its impact on IQ of schoolchildren in Mundra, Gujarat, India",
Trivedi et al, Fluoride 45(4)377–383 October-December 2012, http://fluoridealert.org/uploads/trivedi-2012.pdf
115 "Effect of fluoride exposure on the intelligence of school children", Saxena et al, J Neurosci Rural Pract. 2012 May-Aug; 3(2): 144–149. doi: 10.4103/0976-3147.98213, http://www.ncbi.nlm.nih.gov/pmc/articles/PMC3409983/
116 See Environmental exposure to lead and children's intelligence at the age of seven years. The Port Pirie Cohort Study", Baghurst et al, N Engl J Med. 1992 Oct 29;327(18):1279-84. And also see "Low-level lead exposure and children's intelligence from recent epidemiological studies in the U.S.A", Koike S., Nihon Eiseigaku Zasshi. 1997 Oct;52(3):552-61. http://www.ncbi.nlm.nih.gov/pubmed/9388360
117 "Effect of High Water Fluoride Concentration on the Intellectual Development of Children in Makoo/Iran",
B. Seraj et al, Journal of Dentistry (Tehran). 2012 Summer; 9(3): 221–229. http://www.ncbi.nlm.nih.gov/pmc/articles/PMC3484826/

"Children residing in areas with higher than normal water fluoride levels demonstrated more impaired development of intelligence. Thus, children's intelligence may be affected by high water fluoride levels."

Although all water contains a natural background level of fluoride, usually between 0.1ppm and 0.3ppm in NZ, which is unavoidable, the studies suggest there may be no 'safe' level of fluoride: it dumbs you down, end of story. Dosing drinking water to a full 1ppm, with an extra dose above that again if you use fluoride toothpaste twice daily, appears to be begging for trouble. Throw in a few cups of tea each day (very high in fluoride[118]) and you could well and truly pay for it down the track.

One of the Codex advisors is New Zealander Peter Cressey. In 2009 he investigated fluoride ingestion in New Zealand, and discovered that babies fed infant formula using fluoridated water at 1 part per million were being over-fluoridated: "at a water fluoride concentration of 1.0 mg/L the Upper Limit would be exceeded 93% of the time."[119]

Given the impact of fluoride on IQ, the news that babies are being regularly overdosed on the toxic chemical in their first year of life is not good. The New Zealand study did not look at the compounding impact on brain chemistry of fluoride and aluminium taken together.

A major review of available studies last year and published on the US PubMed database backs up everything you've just read: fluoride is a direct cause of lower IQ in children and may significantly harm the future mental capacity of unborn children.[120]

How could we possibly have a situation where health authorities in New Zealand, Australia, the US, UK and Canada fall over themselves to impose population-scale medication with fluoride, when study after study shows fluoride makes people less intelligent?

Great question, and the answer may actually be buried in that question if you look at it objectively. You probably won't be surprised to hear fluoride is another one of those "special" additives:

"Fluoride compounds which are put: in water (fluoridation), in toothpastes, in supplemental tablets were never tested for safety before approval,"

118 http://www.health.govt.nz/our-work/preventative-health-wellness/fluoridation/fluoride-sources
119 http://www.esr.cri.nz/SiteCollectionDocuments/ESR/PDF/MoHReports/FW0651-Fluoride-intake-assessment-July2009.pdf
120 "Developmental Fluoride Neurotoxicity: A Systematic Review and Meta-Analysis", Choi et al, Environ Health Perspect. 2012 October; 120(10): 1362–1368. doi: 10.1289/ehp.1104912, http://www.ncbi.nlm.nih.gov/pmc/articles/PMC3491930/

records a Romanian government study on the history of fluoridation.[121]

The science also shows that the impact of fluoride on brain degeneration is magnified by the presence of aluminium: "Fluoride alone, or in combination with aluminium, triggers a cascade of molecular events," reports one recent study.[122] Knowing that the chemical companies are pumping aluminium into processed foods, and that fluoride is a chemical waste product being sold to taxpayers, again you can start to join the dots by following the money.

There's one more major study we should review on fluoride. In 2013, the *European Journal of General Dentistry* carried out a thorough analysis of the fluoride debate, asking whether the chemical still deserved a place in "preventive dentistry". You will discover their answer in a moment, but first here's what they discovered on the way.[123]

"Fluoridation was first introduced as a public health measure in the USA in the 1950s, after cross-sectional studies of naturally fluoridated regions of that country suggested that levels of tooth decay declined as the fluoride concentration in drinking water increased. Several 'controlled fluoridation trials' were conducted in the USA and Canada.

"Then, in Australia, the National Health and Medical Research Council, Australian Dental Association and Australian Medical Association all endorsed fluoridation in the 1950s, despite considerable opposition from doctors in the letters columns of the *Medical Journal of Australia*. At that time there was almost no knowledge of the mechanisms of action of fluoride in the human body."[124]

In other words, as we confirmed above, there was no safety testing carried out on fluoride before it was thrown into public water supplies. The study continues:

"Despite the presence of enormous data on the beneficial effects of fluorides in prevention of dental caries the fact that fluoride has serious adverse effects can not be ignored. The most common side effect noticed is dental fluorosis which occurs as a result of fluoride overdose and results

121 "Fluoride – the Danger that we must Avoid", H. BÅLAN, ROM. J. INTERN. MED., 2012, 50, 1, 61–69, http://www.intmed.ro/attach/rjim/2012/rjim112/art08.pdf
122 "Neurotoxic Effects of Fluoride in Endemic Skeletal Fluorosis and in Experimental Chronic Fluoride Toxicity",
Shivarajashankara Y.M, Journal of Clinical and Diagnostic Research, 2012, May, Vol: 6 Issue 4, 740 – 744
123 "Is fluoride still a pivot of preventive dentistry?" Mahajan et al, European Journal of General Dentistry, Feb 2013 Volume 2 Issue : 1 Page : 20-24, DOI: 10.4103/2278-9626.106797, http://www.ejgd.org/article.asp?issn=2278-9626;year=2013;volume=2;issue=1;spage=20;epage=24;aulast=Mahajan#ft30
124 Diesendorf M. A kick in the teeth for scientific debate. Australasian Science 2003;24:,35-37

in tooth discoloration a condition called "mottled enamel".

"In artificially fluoridated regions, dental fluorosis is now much more prevalent and severe than the initial proponents of fluoridation predicted. The University of York's Fluoridation Review,[125] [8] estimates that *up to 48% of children in fluoridated areas have some form of dental fluorosis.* [emphasis added]

"In addition to dental fluorosis there is also a large and growing body of research on a fluoride-induced bone disease called skeletal fluorosis. This disease is observed on X-rays as increased bone density, structural damage to bones, and calcification of joints and ligaments. In severe cases, some patients cannot even straighten their arms or even walk upright."

Take home points so far: yes, fluoride in the water reduces dental cavities (to a limited extent, as you will shortly see), but the dangers of fluorosis caused by exposing children to fluoridated water are much more serious than originally expected. Up to half the children with fluoridated water supplies now exhibit symptoms of fluorosis.

The study continues:

"Most people assume that these severe manifestations of skeletal fluorosis occur at much higher fluoride levels than the 1 part per million. To the contrary, clinically significant cases of skeletal fluorosis have been reported in at least 9 papers from 5 countries...A few cases are even reported in India and China at fluoride concentrations slightly below 1 ppm."

That's because we get fluoride from more than just drinking water, making water dosing a dangerous gamble. The study continues:

"Some worrying results have also been published on the biological effects of fluorides, based on laboratory and animal experiments. It is well known to biochemists that, contrary to one of the pro fluoridation myths, fluoride is highly active biologically, forming a strong hydrogen bond with the groups found in proteins and nucleic acids.[126] In vitro experiments demonstrated that fluoride inhibits enzymes, and induces chromosome aberrations,[127] and genetic mutations.[128]

125 Treasure ET, Chestnutt IG, Whiting P, McDonagh M, Wilson P, Kleijnen J. The York review – a systematic review of public water fluoridation: A commentary. British Dental Journal 2002;192:495-7. http://www.ncbi.nlm.nih.gov/pubmed/12047121?dopt=Abstract

126 Emsley J, Jones DJ, Miller JM, Overill RE, Waddilove RA. An unexpectedly strong hydrogen bond: A initio calculations and spectroscopic studies of amide-fluoride systems. J Am Chem Soc 1981;103:24-8.

127 Suzuki N, Tsutsui T. Dependence of lethality and incidence of chromosome aberrations induced by treatment of synchronized human diploid fibroblasts with sodium fluoride on different periods of the cell cycle. Shigaku1989;77:436-47

128 Caspary WJ, Myhr B, Bowers L, McGregor D, Riach C, Brown A. Mutagenic activity of fluorides in mouse lymphoma cells. Mutat Res 1987;187:165-80.

"Professor Anna Strunecka of Charles University in the Czech Republic has shown in laboratory experiments that fluoride in the presence of aluminum disrupts G-proteins.[129] G-proteins take part in a wide variety of biological signaling systems, helping to control almost all important life processes. Furthermore, pharmacologists estimate that up to 60% of all medicines used today exert their effects through a G-protein signaling pathway.

"Animal experiments reveal that fluoride increases the uptake of aluminum into the brain at 1 ppm in the drinking water...Thus, the malfunctioning of G-proteins could be a causal factor in many human diseases, including Alzheimer's disease, asthma, memory disturbance, migraine and mental disorders."[130]

Now, the important thing to remember is that this study was published 2013 in the *European Journal of General Dentistry*. This is a medical journal for dentists. Which makes its conclusion stunning:

"Fluoride because of its anti-caries action was considered a pivot of preventive dentistry. It was considered as a double edge sword as the excess amount was responsible for dental as well as skeletal fluorosis, which is incurable. But its benefits as an anti-caries element were so much endorsed that it over shadowed its serious side effects. But with a changing scenario attention is now being drawn on potentially permanent damaging effect of fluoride.

"This review of literature on fluoride research reveals a situation where people in fluoridated communities are required to ingest a harmful and ineffective medication with uncontrolled dose. The medication actually doesn't need to be swallowed, since it acts directly on tooth surfaces. *The benefit of fluoridation is at best a reduction in tooth decay in only a fraction of one tooth surface per child*. [emphasis added][131]

"It is time for advanced nations and fluoridating countries to recognize that fluoridation is outdated and has serious risks that far outweigh any minor benefits, violates sound medical ethics and denies freedom of

129 Strunecka A, Patocka J.Pharmacological and toxicological effects of aluminofluoride complexes. Fluoride 1999;32:230-42
130 Varner JA, Jensen KF, Horvath W, Isaacson RL. Chronic administration of aluminum-fluoride and sodium-fluoride to rats in drinking water: Alterations in neuronal and cerebrovascular integrity, Brain Res 1998;784:284-98
131 See Brunelle JA, Carlos JP. Recent trends in dental caries in U.S. children and the effect of water fluoridation. J Dent Res 1990;69:723-7., and also Spencer AJ, Slade GD, Davies M. Water fluoridation in Australia. Community Dent Health 1996;13:27-37, http://www.ncbi.nlm.nih.gov/pubmed/8897748?dopt=Abstract

choice. With the advancement of recent methods for caries prevention role of fluoride in preventive dentistry needs to be readdressed."

Explaining its position, the *European Journal* article explained that the pro-fluoridation campaigns run through health ministries appear to have been captured by powerful "corporate interests":

"Behind the dental and medical associations, who promote fluoridation, are powerful corporate interests like-the sugary food industry (e.g., sugar, soft drinks, processed breakfast cereals and sweets) that benefits from the notion that there is a magic bullet that stops tooth decay, whatever junk food our children eat; the phosphate fertilizer industry that sells its waste silico-fluoride to be put in drinking water instead of paying for its safe disposal; and the aluminum industry, which had an image problem with the atmospheric fluoride pollution it produces, and funded some of the early research in naturally fluoridated regions of the USA that appeared to show that fluoride was good for teeth.

"Some governments support fluoridation because they consider it to be a cheaper way of addressing tooth decay than running effective dental services for school children and older people, and politically safer than tackling the promotion of sugary foods that are the main cause of tooth decay."

If you read the literature of the Club of Rome, Ted Turner or some of the other totalitariacrats, one of their policies is the urgent need for depopulation. Funnily enough, fluoride appears to assist in that process. One of the primary causes of death in the elderly is hip fracture.

In 1992 researchers reported a "significant increase"[132] in the rates of hip fractures in fluoridated areas, because fluoride attacks bones. A subsequent study found "about one-half of the people with hip fractures end up in nursing homes, and in the year following the fracture, 20 per cent of them die," confirmed Harold Slavkin, Director of National Institute of Dental Research, to the *Journal of the American Dental Association* in 1999.

A report down-under reached the same kind of conclusions, noting that five major studies in the US, UK and France had confirmed much higher rates of hip fractures in fluoridated areas.[133]

Hip fractures are often a death sentence for older people. It appears they are avoidable.

Hopefully, this chapter has clarified the linkages between big pharma-

---

ceutical companies and their fraudulent scientific testing, big chemical companies, food additives and public health campaigns based on industrial toxic waste.

You'll be pleased to know that these global giants are among those driving for global free trade agreements like the Trans Pacific and Trans Atlantic partnerships. These corporates are literally just a few steps away from effective financial control of the world economies.

CHAPTER 8

Controlling Trade

*"Free trade is the principal point in the program
of internationalism."*
– Christian Lous Lange, Norwegian Globalist

In 2005, little old New Zealand, home of the close-run America's Cup challenge, initiated a free trade deal with three other Asia-Pacific rim states: Singapore, Brunei and Chile. Sensing that the 21st century would be dominated by the Asia Pacific rim, not Europe, the USA under the Bush administration began tyre-kicking the free trade deal and in early 2008 commenced negotiations to join. Once America's interest became plain, other countries scrambled for a seat at the table as well: Australia, Peru, Vietnam, Malaysia, Mexico, Canada, Japan and Taiwan.

Between them, these countries control somewhere between 30 and 40% of all world trade. As the first country in the world to negotiate a free trade deal with China, New Zealand was in the box seat to drive the Trans Pacific Partnership deal.

At the same time, the United States has now opened negotiations with Europe for a Trans Atlantic Free Trade Agreement as well.

The key thing about all these negotiations however, is the secrecy that surrounds them. Under international law, delegates to treaty negotiations have power to approve the text, and politicians have the power to sign off on it. Once the treaty is signed off and ratified, it is binding forever unless there's an out clause.

We don't know whether there's an out clause because the treaty drafts have never been officially released, and only a few select pages have been leaked to the media.

"Six hundred US corporate advisors have had input into the TPP. The draft text has not been made available to the public, press or policy makers. The level of secrecy around this agreement is unparalleled," reported one anti-TPP media lobby group.[134] "The majority of Congress is being kept in the dark while representatives of US corporations are being consulted and privy to the details.

"The chief agricultural negotiator for the US is the former Monsanto lobbyist, Islam Siddique. If ratified the TPP would impose punishing regulations that give multinational corporations unprecedented right to demand taxpayer compensation for policies that corporations deem a barrier to their profits.

There appears not to be a specific agricultural chapter in the TPP. Instead, rules affecting food systems and food safety are woven throughout the text. This agreement is attempting to establish corporations' rights to skirt domestic courts and laws and sue governments directly with taxpayers paying compensation and fines directly from the treasury."

Politicians, like New Zealand trade minister Tim Groser, have accused protestors of scaremongering in an attempt to "wreck" the agreement. Unfortunately there is no way of knowing, because the New Zealand and American governments are refusing to release the draft treaty texts. The public will not find out what they're up to until the moment after the deal has been signed.

The concept of multinational companies being able to sue national governments for breach of any treaty conditions is logical, however. The multinational corporates are all over the TPP like a rash because the point of the deal is to give them free access to markets, and that free access has to be backed up by legal sanctions.

What we do know from the leaks so far is that the TPP will forbid countries from imposing food labelling requirements on Genetically-modified food. That means consumers will not be allowed to know whether the food they are eating contains GM ingredients.

Citing the effect of the ten year old NAFTA free trade agreement on Mexico, the lobbyists note that "Monsanto, DuPont and Syngenta now

---

134 http://www.nationofchange.org/trans-pacific-partnership-and-monsanto-1372074730

control 57 percent of the commercial food market" in Mexico.[135]

The *New York Times* has weighed in as well, calling the TPP "the most significant international commercial agreement since the creation of the World Trade Organization in 1995".[136]

The agreement, it says, "would set new rules for everything from food safety and financial markets to medicine prices and Internet freedom."

In other words, it will control your life.

Members of the US Congress have not been allowed access to the whole TPP draft, but one Democrat congressman, Alan Grayson, has seen some of it, on the condition he did not reveal what he saw. Afterwards, he fumed to the *Huffington Post*:

"Having seen what I've seen, I would characterize this as a gross abrogation of American sovereignty," Grayson told *HuffPost*. "And I would further characterize it as a punch in the face to the middle class of America. I think that's fair to say from what I've seen so far. But I'm not allowed to tell you why!"[137]

If only you knew. There are two ways of achieving a global administration. One is from the top down, and the other is from the bottom up. Trying to impose a global solution from the top down is asking for trouble. But creating the conditions worldwide that make a global form of government inevitable – that's easy. It might take a little longer, but it works on the boiling frog principle. By the time the public wake up to what's happening, most of it has already happened bar the shouting.

It's incredible how long those who dreamed this up have been working to see it come to fruition, as we are about to find out.

---

135 Ibid
136 "Obama's Covert Trade Deal", by Lori Wallach and Ben Beachy, New York Times 2 June 2013, http://www.nytimes.com/2013/06/03/opinion/obamas-covert-trade-deal.html?_r=2&
137 "Alan Grayson On Trans-Pacific Partnership: Obama Secrecy Hides 'Assault On Democratic Government'", Huffington Post, 18 June 2013, http://www.huffingtonpost.com/2013/06/18/alan-grayson-trans-pacific-partnership_n_3456167.html

CHAPTER 9

Controlling Society Through Migration

"A nation that cannot control its borders
is not a nation."
– Ronald Reagan, US President

It has been said in this book, and probably will be again, that the road to Hell is paved with good intentions. As a loose translation, it means people and groups can do things for essentially the right reason, but without realising the damage they may eventually cause. Naturally there are those who also do things simply to make a lot of money and gain a lot of power.

The debate about "free trade" is not new. We've been having it for nearly 200 years

It was David Rockefeller who said the future of the world economy would depend on free trade and free movement of peoples in a world without borders. That is, of course, a very simplistic rendition of the problem. To understand why he said it, you first have to understand a little more about the world as it existed, and the world as Rockefeller idealises it.

In the world a hundred and fifty years ago, most markets were domestic. The consumer lived in a small village, a local butcher provided meat, the baker bread, and so on along the food chain. Yes, there was international trade in luxury items like sugar and spices, or timber – which Britain had long since run out of for ship masts and was raiding forests in Virginia and New Zealand – but the vast bulk of commerce was highly localised to nearby communities.

The opening up of railway lines in Europe, Britain and North America in the 1800s caused an explosion of economic growth. Much bigger cargoes could be hauled by trains than horses, and faster. Trains put new frontiers within easy reach, bringing migrants and pioneers to new areas. This wasn't such a major issue in either Britain or Europe, where all available land had long been settled. In America, however, railroads were the making of the west.

With industrialisation and migration came, as always, opportunity. In his day, David Rockefeller's grandfather John D. Rockefeller was his generation's Steve Jobs or Bill Gates. More so, in fact, because by the time he died just shy of his 98th birthday in 1937, John D.'s fortune was around one and a half percent of total US GDP, making him the richest American, in real terms, to have ever lived.

John, born 1839, made his money as America's first oil magnate. Black gold, Pennsylvania tea, and his company Standard Oil went on to become the world's largest petroleum monopoly until it was finally broken into smaller competing business units.

Just as there were oil monopolies in the nineteenth century, so too were there railroad monopolies and other sectors where companies swallowed up all the competition. In fact, if you leave the market to itself, eventually monopolies will emerge. This is neither a good thing nor a bad thing, it just is.

The same principle applies in nature: apex predators pretty much control their own hunting ground. In nature, however, apex predators are not protected from the consequences of their actions by the rule of law. It is a peculiarity of human affairs that big business can price gouge the public by virtue of a monopoly position, but the intended victim is not then permitted to turn around and shoot the CEO of the offending monopoly.

That's why we now have a smattering of laws prohibiting anti-competitive trade behaviour, to try and restore some of the balance. In business then, beginning with the Rockefellers, we have recognised that too much power concentrated in too few hands, without the balance of genuine competition, is a bad thing.

So why on earth are we proposing to go down that path in politics and globalisation all over again?

For the Rockefellers there was truth and reality in the statement that the world economy depends on free trade and free movement of labour. Business, which is a commercial rather than moral endeavour, runs more efficiently with free trade. The ability to contract out labour to workers in cheaper countries, or alternatively to import cheaper migrant workers to staff your factories, is an ideal that big business aims for.

Small business, the butchers, bakers and candlestick-makers of modern times, doesn't need free movement of peoples across borders because by and large they're not big enough to take advantage of such economies of scale. A family who own a corner store, for example, won't be randomly importing cheap labour from overseas because the whole point of their store is to provide income and job-sharing opportunities for their own family.

SMEs, or small to medium enterprises, are the backbone of western economies. They can be found in leased offices or in the back-rooms of people's homes. They employ anywhere from one to 25 people, and they account for most of the paid workforce outside the State's public service. Free trade can be useful to them in terms of keeping import costs low, or opening up small niche markets overseas, but the bread and butter of their income is usually generated in the local economy. SME's may aspire to think globally, but they begin by acting locally.

That's not to say there are not some small businesses who have found international markets for their work. There are people working in New Zealand as freelance contractors for a British legal transcription service, for example. While head office staff in the UK are asleep, on the other side of the world contractors are typing up legal documents that will be ready first thing in the next UK morning.

Getting skilled people to work the graveyard shift in Britain would be harder and probably more expensive. It makes economic sense and an efficient use of timezones to share this work across borders.

Big business, in contrast, is an SME that has hit critical mass at some point. It has become too big for its home market, and expanded into other regions or even other countries. Many big businesses have become global brands, and their input and campaign donations are frequently sought by politicians. Today's Microsoft was yesterday's Dutch East India Company.

This is the situation we have come to in the space of two hundred years. The opening up of the American west, the opening up of world trade routes and travel, coupled with the arrival of air transport and electronic communication, literally re-created a 'Tower of Babel' scenario. For the first time in history, virtually all of humanity is interconnected. English is becoming the common language of the world, and politically where we once had parliaments in our nations – now we have added a parliament *of* nations, in the form of the UN.

Where once nation-states stood, fiercely protective of their own communities and commercial interests, now modern business and modern technology has brought with it a pan-global worldview: the entire world is a 'market', the free flow of labour and contracts is a must.

But is it?

This is the argument at the centre of the globalisation debate? Should we sacrifice our distinct cultures and work environments on the altar of globalisation? Should we be forced to compete not just against the person next door paying a similar rent and grocery bill, but against someone in another country living in a hovel and being paid fifty cents an hour to do the same kind of work?

That's the tricky part about globalisation theory. Big business can afford to set up in different countries to take advantage of such economic "benefits". Ordinary smaller businesses can't. So what happens? Smaller enterprises become less profitable over time and are either driven by cheap competition into the ground, or bought out by their larger competitors. Gradually, more monopolies or cartels build, and economic power consolidates in the top tiers.

Those businesses become so big that it becomes impossible to compete with them. An independent department store, for example, cannot muster the kind of bulk purchasing discounts that a chain with a thousand stores can. And if the independent store slashes its prices at its one location, the chain store can maintain its overall profitability by competing head on in that location or even undercutting, while cross-subsidising its losses there with profits from other areas where no competition exists.

John D. Rockefeller knew a thing or two about eliminating competition. His grandson David Rockefeller knows those things too. There is a belief in a kind of benign utopia where people at the top of the tree make decisions for the good of everyone else. It's a belief that development can be and should be guided by people who know best.

In this respect, the capitalist right represented by the Rockefellers and others has entered into a marriage of convenience with elements of the Marxist left, who also see advantages in a managed economy. And I'm not the first to make that observation.

Professor Carroll Quigley taught history at Georgetown University in the US. He taught future president Bill Clinton. In the 1960s he wrote a 1,300 page book entitled *Tragedy and Hope*, based on his life's work investigating the history of the preceding century. He wrote that world affairs were being guided by an 'elite':

"There does exist, and has existed for a generation, an international Anglophile network which operates, to some extent, in the way the Radical right believes the Communists act. In fact, this network, which we may identify as the Round Table Groups, has no aversion to cooperating with the Communists, or any other group, and frequently does so. I know of the operation of this network because I have studied it for twenty years and was permitted for two years, in the early 1960s, to examine its papers and secret records.

"I have no aversion to it or to most of its aims and have, for much of my life, been close to it and to many of its instruments. I have objected, both in the past and recently, to a few of its policies... but in general my chief difference of opinion is that it wishes to remain unknown, and I believe its role in history is significant enough to be known."

Quigley identified what he believed was the first public manifestation of the secret network in the 1890s, and named scholar and explorer Cecil Rhodes as one of its instigators. Rhodes was indeed a globalist, and he was also a Freemason.[138] He believed he could instigate a New World Order under the joint control of the US, UK and Germany. Quigley documents in his book how Rhodes quickly gathered like-minded thinkers around him and they proceeded to use donations of wealth to influence education policy at the world's major universities.

"The power and influence of this Rhodes-Milner group in British imperial affairs and in foreign policy since 1889, although not widely recognised, can hardly be exaggerated," the Georgetown professor wrote in 1966. The group "dominated" the *Times of London* newspaper and "controlled it completely" between 1912 and 1966. Britain's leading newspaper was the mouthpiece of Rhodes' agenda.[139]

Numerous other 'papers and journals' subsequently came into the orbit of Rhodes' followers. "They have also established and influenced numerous university and other chairs of imperial affairs and international relations."

Not to mention the ubiquitous Rhodes Scholarship programme for future world leaders. It was the Rhodes group, he writes, that transformed the British Empire into what we now know as the Commonwealth, in a bid to

---

138 See http://www.apollo357.com/index.php/history/1870-1914 and in particular details of Cecil Rhodes funeral towards the end of the document.
139 Tragedy & Hope by Carroll Quigley, MacMillan New York, 1966, p133

prepare for a global governmental system run out of either London (first choice) or Washington, organised "in a federal system". The joining of "the five colonies in Australia" into one nation in 1901 and the unification of South Africa in 1910 were two items on the bucket list that got ticked off.

One of Rhodes' disciples, Lionel Curtis, had, writes Quigley, "a fanatical conviction that with the proper spirit and the proper organisation (local self-government and federalism), the Kingdom of God could be established on earth."[140]

The significance of this apparent spiritual motivation behind the world government movement will become more obvious as you read through this book.

The Rhodes philosophy was all about uniting the world under one civilisational banner. In this new world, there would be free movement of peoples as citizens of a common government.

The debate about removing national borders to allow free movement of labour has been carefully propagandised to become a debate about multiculturalism and racism. It's no longer possible to have a genuine debate about the importation of labour because vested interests turn it into a race issue virtually every time.

Over the past two centuries, America in particular has welcomed many migrants from different countries. Until recently, you could argue that this was still a form of monoculturalism in that the Irish, Italians, Danes, Spaniards, Poles, Germans and others who purchased passage to the New World were still from a Judeo-Christian culture: we may not speak the same language or observe the same holidays, but broadly-speaking we share the same values.

It would be called "xenophobia" today, but US President Teddy Roosevelt made a speech welcoming migrants to America, as long as they became true Americans:[141]

"In the first place, we should insist that if the immigrant who comes here in good faith becomes an American and assimilates himself to us, he shall be treated on an exact equality with everyone else, for it is an outrage to discriminate against any such man because of creed, or birthplace, or origin. But this is predicated upon the person's becoming in every facet an American, and nothing but an American.

"There can be no divided allegiance here. Any man who says he is an American, but something else also, isn't an American at all. We have room

---

140 Ibid, p 146
141 http://www.snopes.com/politics/quotes/troosevelt.asp

for but one flag, the American flag... We have room for but one language here, and that is the English language... and we have room for but one sole loyalty and that is a loyalty to the American people."

You would never get a government politician to say such a thing today, because we have been conditioned to abhor nationalism. Roosevelt was not saying discriminate, he was saying "when in Rome, do as the Romans do." Nowadays, the West has been taught to self-loathe its culture.

Post WWII, there's been a much greater intake of refugees and migrants from cultures other than European. This is where the marriage of convenience between Left and Right comes in. Big business wanted cheap labour. The Left wanted to spread the wealth and lift up the Third World by giving them access to western economies.

Adding fuel to this mix is a massive drop in birth rates in Western countries. The native cultures in most European and Asian countries are failing to meet population replacement levels in the balance between births and deaths. Replacement level is 2.1 children per woman, but in Japan and parts of Europe birth rates have fallen as low as 1.4.

The reasons for this are many. Contraception is freely available. Jobs have been restructured so that the salaries once large enough to support one breadwinner are now spread across two, creating greater productivity but with the bonus for corporates of lowering real wages because more people are seeking work, driving the overall cost of labour down. By the time many women have built up sufficient capital in their career that they feel comfortable starting a family, they have left it so late that fertility becomes an issue.

Abortion is another reason for low birth rates in the west. Around one in every three pregnancies ends in the baby's death. A staggering 800 million potential wage-earners and taxpayers have been killed in the womb worldwide since the Roe v Wade decision legalising abortion in 1973.

Now, you could argue that the world doesn't need 800 million more people, but the reality is we haven't actually been dropping our populations in the west, we've killed our own progeny but made up the numbers bringing in someone else's kids.

Here's an example. In New Zealand, somewhere in the region of 19,000 abortions take place each year. But the country has a skilled labour shortage. New Zealand is welcoming up to 60,000 foreign born migrants every year.[142] Whatever the argument you try and make for abortion on

---

142 http://www.dol.govt.nz/publications/research/plt-migration-big-picture/03.asp

the grounds of overpopulation, it simply is not valid in any country that also allows mass immigration.

The West has become increasingly reliant on inflows of cheap labour – often from third world nations – to provide fodder for economic growth, and to provide a younger demographic as the native cultures with their low birth rates age.

The impact in New Zealand is huge. Official statistics show that of 3.4 million New Zealand adults in 2010, a staggering 1.6 million were either immigrants or first generation New Zealanders (children of a migrant). Only 1.76 million adults were second generation New Zealanders or longer.[143]

Mass migration, particularly when coupled with abortion, is a form of social engineering being practiced on a global scale.

"American military historian and columnist Victor Davis Hanson talks about how mass immigration is the product of a de facto alliance between the Libertarian Right and the Multicultural Left," notes an article in the European journal *Brussells Journal*.

Immigration is a hot-button issue in Europe, where most of the new arrivals have come from Africa, the Middle East or South Asia. Predominantly they are Muslim. Unlike past interactions between similar cultures, modern migration has taken a colour-blind and creed-blind approach, with officials assuming no difference between cultures.

Of course, the reality is different. Most of humanity's big wars have been fought between rival cultures, and the history of clashes between Islam and the West spans some 1,300 years of conflict. Muslims take their faith much more seriously than most in the liberal west do, making flashpoints almost unavoidable.

It is deliberate, from the view of policy planners. By forcing rival cultures to mix they hoped to knock the rough edges off each culture and impose a doctrine of tolerance that would enable movement to the next level of integration.

Swedish libertarian writer Johan Norberg, author of "In Defence Of Global Capitalism", claims the West must have immigrants or our civilisation will die.

"At the moment there is a problem. The right supports one part of globalisation – the free movement of capital and goods – while the left

---

143 http://nzdotstat.stats.govt.nz/wbos/Index.aspx?DataSetCode=TABLECODE7930#

tends to support another part, the free movement of people.

"The big question is whether business and welfare systems will be able to keep going if we don't step up immigration. The UNFPA (UN Population Fund) estimates that the EU will need 13.5 million immigrants a year to keep the ratio between the working and the retired populations constant for the next 50 years. Immigrants aren't a burden, they're an asset, and even refugees put more into the national treasury in the course of a lifetime than they get out of it. If large numbers of migrants become permanently dependent on handouts, that's merely an argument for seriously reforming our social security systems and our labour market regulations. But that isn't the main argument. The main argument is that being allowed to migrate, even if there is a national boundary in the way, is in fact a human right. The western world moralised, and rightly so, about the communists forbidding their citizens to leave their native country. But now that they are allowed to do so, we are forbidding them to enter our countries."[144]

That's Norberg's opinion, but the reality remains: if you are killing your own future citizens and importing someone else's citizens to fill the gap, you can hardly complain when people whose allegiance is to a different culture eventually call the shots in your country. In a truly freemarket world with no borders, eventually there are no countries, only homelands – like the sixth generation Americans who still hearken back to their Irish roots. Ironically, if Norberg's vision of utterly open borders comes to pass, our civilisation may die regardless.

The *Brussells Journal* takes Norberg to task.

"Norberg can have valuable insights into the flaws of the Scandinavian welfare state model. However, his commitment to a 'free market, open border' ideology blinds him to the threat posed by Muslim immigration, an ideological blind spot that is almost as big as the ones we find in Marxists.

"Norberg believes immigration is already so extensive it would be unwise to halt it. Pointing out there were 15 million Muslims in Europe, he noted in a 2003 article: 'If we close the borders, if we alienate this substantial minority, we risk creating resentment between ethnic and religious groups, and only the fundamentalists would gain...If people were allowed to cross borders at will, they would take their ideas and their labour and skills with them. This is all part of free trade, and it's a

---

144 http://www.johannorberg.net/?page=indefense_qa

paradox that many liberals don't see this'."

One liberal who definitely saw the advantages was Karl Marx:

"Under the freedom of trade the whole severity of the laws of political economy will be applied to the working classes. Is that to say that we are against Free Trade? No, we are for Free Trade, because by Free Trade all economical laws, with their most astounding contradictions, will act upon a larger scale, upon a greater extent of territory, upon the territory of the whole earth; and because from the uniting of all these contradictions into a single group, where they stand face to face, will result the struggle which will itself eventuate in the emancipation of the proletarians."[145]

In other words, free trade and free movement would create the social conditions necessary for a worldwide revolution. Ironic that Communism and Capitalism have jointly led us to this point.

Marx expanded on the theme:

"But, in general, the protective system of our day is conservative, while the free trade system is destructive. It breaks up old nationalities and pushes the antagonism of the proletariat and the bourgeoisie to the extreme point. In a word, the free trade system hastens the social revolution. It is in this revolutionary sense alone, gentlemen, that I vote in favour of free trade."[146]

Make no mistake, the totalitariacrats pushing mass migration and free trade agreements are university educated. They know all about Marxian theory. They know full well where they are taking you.

It's not that migration and free trade are inherently evil – they're not. But like many things in life, too much taken too fast can be fatal. What's interesting is that even back in 1848 the spin doctors were using techniques later put to good use in the climate change propaganda war.

"They send an army of missionaries to all corners of England to preach the gospel of free trade," wrote Marx. "They have printed and distributed gratis thousands of pamphlets to enlighten the worker upon his own interests, they spend enormous sums to make the press favourable to their cause; they organize a vast administrative system for the conduct of the free trade movement, and they display all their wealth of eloquence at public meetings."[147]

Al Gore, eat your heart out. Telling porkies, and getting the mainstream

---

145 Marx quoted in "The Free Trade Congress at Brussels" by Frederick Engels, *Northern Star*, 1847 http://www.marxists.org/archive/marx/works/1847/09/30.htm

146 "On the Question of Free Trade", 9 January 1848, http://www.marxists.org/archive/marx/works/1848/01/09ft.htm

147 ibid

media to help you hoodwink the public, evidently has a long pedigree.

Remember those statistics a few paragraphs ago indicating a large proportion of New Zealand adults are migrants or first generation New Zealanders? Now take a trip back to Britain in 1968, where Conservative MP Enoch Powell gave a speech that sounded the death knell of his career. At that time, Britain was welcoming migrants from its West Indies and African colonies and official policy was to create a melting-pot kind of society where everyone got along.

Powell, for all his sins, nonetheless saw the cultural tensions building in his own electorate and predicted dark times ahead for Britain if mass migration continued:

"Areas ... are already undergoing the total transformation to which there is no parallel in a thousand years of English history," he said.[148]

"In 15 or 20 years, on present trends, there will be in this country three and a half million Commonwealth immigrants and their descendants. That is not my figure. That is the official figure given to parliament by the spokesman of the Registrar General's Office.

"There is no comparable official figure for the year 2000, but it must be in the region of five to seven million, approximately one-tenth of the whole population, and approaching that of Greater London. Of course, it will not be evenly distributed from Margate to Aberystwyth and from Penzance to Aberdeen. Whole areas, towns and parts of towns across England will be occupied by sections of the immigrant and immigrant-descended population.

"As time goes on, the proportion of this total who are immigrant descendants, those born in England, who arrived here by exactly the same route as the rest of us, will rapidly increase. Already by 1985 the native-born would constitute the majority. It is this fact which creates the extreme urgency of action now, of just that kind of action which is hardest for politicians to take, action where the difficulties lie in the present but the evils to be prevented or minimised lie several parliaments ahead."

As of the 2011 Census, 7.5 million British residents, out of a total population of 63 million people, were born outside Britain. That figure does not include their children (first generation Brits) or their grandchildren (second generation). That is only the number of living migrants now in the UK.[149]

"Those whom the gods wish to destroy, they first make mad," continued

148 http://www.telegraph.co.uk/comment/3643823/Enoch-Powells-Rivers-of-Blood-speech.html
149 http://www.ons.gov.uk/ons/interactive/census-3-1 - -country-of-birth/index.html

Enoch Powell in 1968. "We must be mad, literally mad, as a nation to be permitting the annual inflow of some 50,000 dependants, who are for the most part the material of the future growth of the immigrant-descended population. It is like watching a nation busily engaged in heaping up its own funeral pyre."

By 2012, the annual net migration into Britain was 156,000 a year. Much higher than the 1968 level, but on a per capita basis nowhere near New Zealand's net migration levels of up to 50,000 a year on a population base less than a tenth of Britain's.

When you read Powell's entire speech, it is easy to dismiss him as a racist. At one level, he illustrated the fear of the unknown, and the glaring issue in 1968 Britain was primarily migration from the West Indies by people of "colour". Scratch a little deeper, however, and it's clear the main issues were really differences of culture and economic class, coupled with the natural desire of migrant communities to keep to their own rather than fully integrate. Colour was an easy label, but it was not really the prime cause of dispute.

Powell was protesting the Race Relations Bill, warning that its anti-discrimination provisions would eventually be used to change the very fabric of society in Britain.

"Here is the means of showing that the immigrant communities can organise to consolidate their members, to agitate and campaign against their fellow citizens, and to overawe and dominate the rest with the legal weapons which the ignorant and the ill-informed have provided. As I look ahead, I am filled with foreboding; like the Roman, I seem to see 'the River Tiber foaming with much blood'."

As previously indicated, Powell's future prospects as an MP ended that day. Opposition leader Ted Heath sacked Powell from his shadow Cabinet portfolio, although when Powell died in 1998 Heath remarked that the controversial "Rivers of Blood" speech had been "not without prescience".

Indeed, Powell lived long enough to see the Brixton race riots of the 1980s, but not the beheading of Private Lee Rigby in front of a horrified crowd in central London in May 2013. The emasculated Metropolitan Police – well-known to turn up en masse to a mere speeding ticket – took 20 whole minutes to arrive on the scene, so it was left to passerby Ingrid Loyau-Kennett – a mother and cub scout leader – to talk the jihadist attackers into handing over their weapons.[150]

London streets again ran red with blood in June 2013:[151]

"A 20-year-old man has appeared in court charged with the murder of a young woman who was beheaded in a knife attack.

"Aras Hussein appeared at Sheffield Magistrates' Court today accused of murdering Reema Ramzan, 18.

"Miss Ramzan, who was from the Darnall area of Sheffield, died on June 4 following an incident at a property on Herries Road, in the city.

"Detectives said she suffered a severe knife attack resulting in fatal injuries, including the severing of her head."

And yet again that very same week when 51 year old Daha Mohammed was arrested for trying to behead wheelchair bound amputee Colin Greenaway. Greenaway, who couldn't get away, died in his wheelchair from massive blood loss after his throat was slit open.[152]

Regardless of how "prescient" former Prime Minister Ted Heath regarded Powell's "Rivers of Blood" speech in 1998, it is doubtful that Heath or even Powell himself seriously imagined a time when Britain would experience three beheadings on its streets in the space of less than a month.

Much water has flowed under bridges since those heady multicultural hippy days of 1968, and you can easily see this when you compare Enoch Powell's warnings to the sentiments of current British Prime Minister David Cameron in this speech in 2011 on the problem of homegrown Islamic extremism.[153]

"The root lies in the existence of this extremist ideology. And I would argue an important reason so many young Muslims are drawn to it comes down to a question of identity. What I'm about to say is drawn from the British experience, but I believe there are general lessons for us all.

"In the UK, some young men find it hard to identify with the traditional Islam practised at home by their parents whose customs can seem staid

150 http://www.telegraph.co.uk/news/uknews/terrorism-in-the-uk/10074881/Mum-talked-down-Woolwich-terrorists-who-told-her-We-want-to-start-a-war-in-London-tonight.html
151 "Man in court over beheading of young woman", The Standard, 18 June 2013 http://www.standard.co.uk/news/crime/man-in-court-over-beheading-of-young-woman-8664055.html
152 "Thamesmead murder: neighbour describes sorrow after finding disabled Colin Greenway dead", Bexley Times, 17 June 2013, http://www.bexleytimes.co.uk/news/court-crime/thamesmead_murder_neighbour_describes_sorrow_after_finding_disabled_colin_greenway_dead_1_2239725
153 Speech by UK Prime Minister David Cameron to Munich security conference, 5 February 2011, http://webarchive.nationalarchives.gov.uk/20130109092234/http://number10.gov.uk/news/pms-speech-at-munich-security-conference/

when transplanted to modern Western countries. But they also find it hard to identify with Britain too, because we have allowed the weakening of our collective identity.

"*Under the doctrine of state multiculturalism*, [emphasis added] we have encouraged different cultures to live separate lives, apart from each other and the mainstream. We have failed to provide a vision of society to which they feel they want to belong. We have even tolerated these segregated communities behaving in ways that run counter to our values.

"So when a white person holds objectionable views – racism, for example – we rightly condemn them. But when equally unacceptable views or practices have come from someone who isn't white, we've been too cautious, frankly even fearful, to stand up to them.

"The failure of some to confront the horrors of forced marriage – the practice where some young girls are bullied and sometimes taken abroad to marry someone they don't want to – is a case in point.

"This hands-off tolerance has only served to reinforce the sense that not enough is shared. All this leaves some young Muslims feeling rootless. And the search for something to belong to and believe in can lead them to this extremist ideology.

"For sure, they don't turn into terrorists overnight. What we see is a process of radicalisation. Internet chatrooms are virtual meeting places where attitudes are shared, strengthened and validated.

"In some mosques, preachers of hate can sow misinformation about the plight of Muslims elsewhere. In our communities, groups and organisations led by young, dynamic leaders promote separatism by encouraging Muslims to define themselves solely in terms of their religion. All these interactions engender a sense of community, a substitute for what the wider society has failed to supply."

Britain is running out of time, however. The BBC coyly avoided reporting this year that Muhammed is now the second most popular boy's name in Britain, although it did reveal it was the most common name for a baby boy in London. It was able to avoid revealing the national placement of Muhammed because there are two other variant spellings: Mohammed and Mohammad. None of the three variants on their own break into the top ten boys names, which officially were dominated by "Harry". When their results are added together, however, Muhammed rockets into second place. It was literally only 29 namings behind Harry. If 30 more Muslim women had named their boys Muhammed last year, it would have been number 1 across England and Wales. While the BBC may have failed to mention Muhammed's national prominence, it was left to the *Daily Mail* to reveal all this.[154]

At one level you can argue that no culture exists in isolation, and that we all have to accommodate change. Whilst true for today, though, it has never been the case up until now. Previously, national boundaries in history followed cultural/tribal/racial lines. Britain, France, the Germanic states, they were all largely homogenous societies until the borders came down in recent decades. The only periods of massive change occurred as a result of invasion, and where invasions were successful, as the Normans were in 1066, they brought with them new migrants and changed the culture by force.

The decision by bureaucracies worldwide to fuel their economies with mass migration was a conscious social engineering choice to allow what thousands of years of history had not: a cultural invasion of host countries by peoples formerly and in some cases still hostile to the host countries. In *Star Trek* terms, the *Enterprise* went "shields down" and has remained that way until, as Johan Norberg noted above, the trend has become too big to reverse.

Whoever caused this is almost irrelevant now, the die is cast. Karl Marx must be spinning in his grave like a rotisserie chicken in anticipation of the culture clash now emerging worldwide and its implications for a massive tipping point.

This is not a book lamenting immigration per se. In fact, yours truly is on record from the 1980s arguing the economic benefits of greater migration. What you are reading here is an analysis of the social impact of these policies in the way they have been implemented. It's a classic example, yet again, of divide and rule. The host country does not truly benefit, nor do the disaffected migrant communities, and the social tensions begin to boil over, just as pundits and provocateurs predicted they would.

Gradual, slow migration over decades allows assimilation of migrant communities into the larger social fabric. Rapid waves of migration however encourage the establishment of cultural ghettoes, and those communities become self-sustaining, mini outposts of their homeland, within what to them remains an alien culture.

---

154 "Harry remains most popular boys' name for second year – and Muhammad takes the number two spot", Daily Mail 12 August 2013, http://www.dailymail.co.uk/femail/article-2389817/Harry-Amelia-popular-baby-names-lists-second-year – Muhammad-takes-number-spot.html

Who benefits? These policies have been implemented across the OECD countries. Europe is now internally borderless and has a huge migrant population as well. New Zealand, Australia and Canada all have growing migrant communities. In Australia and Canada localised terror cells have emerged, and New Zealand Prime Minister John Key has reported their existence in New Zealand as well.

The fact that these policies have been initiated globally and are creating havoc at a global level forces us to look a little harder at the context.

Firstly, this is a western problem. There is no massive migrational push into China, Saudi Arabia, Argentina or a range of other ethnically homogenous societies. Instead, it is the West being told it must open its borders to the world, not vice versa.

In New Zealand, which has become the West's de facto canary in the mineshaft on a range of measures, recent national debate has focused on the country's willingness to allow non-resident aliens to buy land and property freely, even though no such reciprocal arrangement exists allowing New Zealand nationals to buy, for example, in China or India.

Although the size of Britain or Japan, the land of New Zealand is home to only 4.5 million people. If only a tiny fraction of the world's population took New Zealand up on its generous policy of selling land to foreign investors, the country would quickly run out of property. Yet this is government policy. The argument has been that we cannot get support for changing the world's structure unless we lead by example.

In America, the official response to eleven million Mexicans storming across the borders is a proposal to award many with citizenship rights if they pay any back taxes that can be proven and pass an exam on being an American.[155]

That might solve the immediate problem of what to do with a large illegal alien problem, but suddenly the solution involves creating 11 million new voters, and then a few years later their children get to vote as well. Suddenly you can see there is political gold being mined.

The totaliatariacrats know that if they throw open the immigration doors, those migrants are likely to be supportive of their initiatives. Divide and rule, carrot and stick, bait and switch. And because, as Karl Marx and others have noted, the social tensions then rise, Big Government can once again ride to the rescue with new "tolerance" laws and more nanny

---

155   http://www.telegraph.co.uk/news/worldnews/us-politics/9876191/Illegal-immigrants-could-be-granted-US-amnesty-within-eight-years.html

state initiatives designed to keep things calm but entrench the system.

In 1986 under Ronald Reagan an alien amnesty was offered, and three million – mostly Mexicans – took it up. It didn't stop illegals from continuing to hop the border:

"I don't think anyone says that it deterred illegal immigration," Cecilia Muñoz, vice president of The National Council of La Raza, America's largest Latino advocacy group told the *Christian Science Monitor.* "But it succeeded in legalizing 3 million people. Their wages went up, and they're fully integrated into American society."[156]

This much, too, is true. Yesterday's migrants have become today's doctors, dentists, entrepreneurs and teachers, on one side of the ledger, and killers, drug dealers and con artists on the other. Just like everyone else, in other words. The debate is not always about the quality of the migrant, it can simply be about the cultural or economic weight that they bring to a host nation.

Again, the essay in the *Brussells Journal* is illuminating:

"We thus have a situation with an explosive population growth in failed countries, while many of the most economically and technologically advanced nations, Eastern and Western, have stagnating populations. This strange and possibly unprecedented situation, which could perhaps be labelled 'survival of the least fit', will have dramatic consequences for the world. It is already producing the largest migration waves in history, threatening to swamp islands of prosperity in a sea of poverty.

"Lenin stated that 'Marxism is based on internationalism or it is nothing'. 'The emancipation of the workers is not a local, nor a national, but an international problem,' wrote Marx. Karl Marx has defined the essence of Socialism as abolishing private property. Let's assume for a moment that a country can be treated as the 'property' of its citizens. Its inhabitants are responsible for creating its infrastructure. They have built its roads and communications, its schools, universities and medical facilities. They have created its political institutions and instilled in its people the mental capacities needed for upholding them. Is it then wrong for the citizens of this country to want to enjoy the benefits of what they have themselves created?

"According to Marxist logic, yes.

"Imagine you have two such houses next to each other. In House A, the

---

156 http://www.csmonitor.com/2006/1106/p13s01-ussc.html

inhabitants have over a period of generations created a tidy and functioning household. They have limited their number of children because they wanted to give all of them a proper education. In House B, the inhabitants live in a dysfunctional household with too many children who have received little higher education. One day they decide to move to their neighbors'. Many of the inhabitants of House A are protesting, but some of them think this might be a good idea. There is room for more people in House A, they say. In addition to this, Amnesty International, the United Nations and others claim that it is 'racist' and 'against international law' for the inhabitants of House A to expel the intruders. Pretty soon, House A has been turned into an overpopulated and dysfunctional household just like House B.

"Among political right-wingers, there is frequently a belief that what is good for business interests is good for the country. The problem is, this isn't always true. There is sometimes a gap between the short-term interests of Big Business for cheap labor from Third World countries, and the long-term interests of the country as a whole. You cannot compete with cheap commodities from Third World countries unless you lower the general wages to Third World levels."[157]

To see a western nation going into cardiac arrest on this issue, return for a moment to David Cameron's Britain and this feature from the *Sunday Times*:

"Just one in six Britons believes immigration has benefited the country, according to one of the biggest studies conducted on the subject. The poll of more than 20,000 people, commissioned by Lord Ashcroft, the former Conservative party deputy chairman, found that 60% of people thought immigration, on balance, had brought more disadvantages than advantages to Britain, with just 17% believing the opposite.

"David Cameron has pledged to reduce net migration from non-EU countries to less than 100,000 a year by 2015, but Ashcroft's research shows politicians have comprehensively failed to win the public's confidence that they are bringing immigration under control.

"Respondents were shown a list of nine suggested measures to cut immigration, such as setting a numerical cap and cracking down on bogus students. Each policy was supported by clear majorities but, in each case, only a minority realised the measure was already government policy.

---

157 http://www.brusselsjournal.com/node/1195

"In another striking finding, 79% of respondents supported the message on the Home Office's 'Go home or face arrest' advertising vans aimed at illegal immigrants. However, just 17% thought the policy would work.

"In his report, to be released tomorrow, Ashcroft says the findings on the vans underline 'the gap' between 'elite and public opinion on immigration'. Writing in today's *Sunday Times*, the peer says: 'Whatever people's view of immigration itself, few think any recent government has had any real grasp of it, or that any of the parties does today'."[158]

When the Conservative Party and its Prime Minister David Cameron are channelling the late Enoch Powell, you know the tipping point has already been reached.

Not to be outdone, one of the UK's top defence chiefs, Rear Admiral Chris Parry, has also compared recent Western developments to the conditions leading up to the fall of Rome:

"Britain and Europe face being overrun by mass migration from the Third World within 30 years, a senior Royal Navy strategist claimed yesterday. In an apocalyptic vision of security dangers, Rear Admiral Chris Parry forecast 'reverse colonisation', where migrants become more dominant than their hosts," the *Daily Mail* reported.

"Globalisation makes assimilation seem redundant and old-fashioned," Parry was quoted, "… [the process] acts as a sort of reverse colonisation, where groups of people are self-contained, going back and forth between their countries, exploiting sophisticated networks and using instant communication on phones and the internet."

As the third world governance systems break down, greater lawlessness and piracy will encroach on modern Europe, he warned.

"At some time in the next 10 years it may not be safe to sail a yacht between Gibraltar and Malta."

"The effects will be magnified," wrote the *Daily Mail* in the article on Parry's comments, "as borders become more porous and some areas sink beyond effective government control. Parry expected the world population to grow to about 8.4 billion in 2035, with some giant metropolises becoming ungovernable. The subsequent mass population movements, Parry argued, could lead to the 'Rome scenario'."[159]

---

158 "Voters say immigration hurts UK", The Times, 1 September 2013 http://www.thesundaytimes.co.uk/sto/public/article1308078.ece
159 "Britain faces mass migration, warns Admiral", Daily Mail, 12 June 2006 http://www.dailymail.co.uk/news/article-390230/Britain-faces-mass-migration-warns-Admiral.html

Interestingly, in New Zealand at least, some of the most vociferous opponents of mass migration and unrestricted foreign investment are recent immigrants. "New Zealand is insane," an Indian caller told national talk host Danny Watson on the Newstalk ZB network in June 2013. "You are letting foreigners buy property and businesses here when New Zealanders cannot go to Mumbai and do the same.

"I have been a New Zealander now for ten years. I love the lifestyle. Don't destroy it."

There are those, however, who want to turn New Zealand, Australia, the US and other western countries into just another federal administration district in a planetary new world order. Fifty shades of vanilla around the world, where everyone is united under one global super-administration working through each national government to get the job done.

In his autobiography, David Rockefeller only touches briefly on his involvement with the Council on Foreign Relations, the Trilateral Commission and the Bilderberg Group. He doesn't deny what is claimed about them. Instead, he says, he is proud:

"Some even believe we [the Rockefeller family] are part of a secret cabal working against the best interests of the United States, characterizing my family and me as 'internationalists' and of conspiring with others around the world to build a more integrated global political and economic structure – -one world, if you will. If that's the charge, I stand guilty, and I am proud of it."[160]

You will read more on the behind-the-scenes push for a world government in coming chapters.

In effect, the new system will mirror the American one, with a world federation composed of individual state governments. It is already underway, and you are about to find out how.

---

160 Memoirs by David Rockefeller, Random House, 2003, p405

CHAPTER 10

Controlling You By International Law

"We are convinced that a comprehensive system of security is at the same time a system of universal law and order ensuring the primacy of international law in politics."
– Mikhail Gorbachev, Soviet Leader, 1987

In the new system, global policy will be set by the super-administration. How each "country" deals with that at a local level will be decided by local politicians, but they will have to work within existing UN treaty obligations (the legal framework) or run the risk of sanctions from the federation.

That's why food laws, climate laws, migration laws, commercial laws and trade laws are all being streamlined under various global agreements.

"Promoting the rule of law at the national and international levels is at the heart of the United Nations' mission," says the UN.[161]

"Establishing respect for the rule of law is fundamental to achieving a durable peace in the aftermath of conflict, to the effective protection of human rights, and to sustained economic progress and development. The principle that everyone – from the individual right up to the State itself – is accountable to laws that are publicly promulgated, equally enforced and independently adjudicated, is a fundamental concept which drives much of the United Nations work."

---

161 http://www.un.org/en/ruleoflaw/index.shtml

By definition, the United Nations cannot make States accountable to a 'rule of law' unless the UN itself, or a similar system, becomes sovereign over those States. That's what the international treaty process is. Signing up to a UN international treaty is a voluntary surrender of sovereignty by the country involved. It is promising under international law to do whatever the treaty says. The treaty under international law has no effect within that country until the local legislature "ratifies" it by passing legislation to do so.

Once the legislation passes, a funny thing happens: your country is bound by the treaty. Yes, your parliament can renounce the treaty at some future point, but it can only do so to the extent permitted by the treaty's own clauses, otherwise it can be penalised with trade and economic sanctions.

For example, many of the major treaties allow for countries to get out, only with the approval of 75% or more of the other treaty partners. That means if other countries want you in, you are in whether your people like it or not.

The European Union financial crisis was a case in point of precisely this system in action. At issue was whether debt-defaulting nations like Greece would be allowed to leave the EU. Up until 2009 it was practically impossible to withdraw from the EU without the consent of the other members of the federation. After 2009 a clause under the Treaty of Lisbon was drawn up permitting a negotiated withdrawal with the approval of other countries.[162]

Unless an international treaty has an explicit clause on withdrawal, however, it is governed by the rules of the Vienna Convention on the Law of Treaties. If your eyes are glazing over, grab a coffee because this is important.

The Vienna Convention came into being in 1969, entered force in 1980 when it reached its quorum, and sets out the international law around treaty obligations. It is a creature of the United Nations, which began work on this back in 1949 and took 20 years to bring it to life. A hundred and thirteen countries around the world have ratified it, including most of Europe, China, Russia, Australia, New Zealand, Canada, Mexico and most of South America and Africa. The United States and Iran have signed it, but not yet passed the Vienna Convention into law.

---

162 http://www.ecb.europa.eu/pub/pdf/scplps/ecblwp10.pdf

Article 54 of the Vienna Convention sets out the rights to withdraw from a treaty:[163]

The termination of a treaty or the withdrawal of a party may take place:

- in conformity with the provisions of the treaty; or
- at any time by consent of <u>all</u> the parties after consultation with the other contracting States.

What this clause means is that a country must either follow the termination procedure spelt out in the treaty text, or if it wants to speed up the process it needs the consent of all other countries who have signed the treaty.

In other words, once the USA ratifies the Vienna Convention, it is bound under international law to honour its treaties regardless of what Congress or the President might feel.

It gets worse, however. Article 56 of the Convention says some treaties cannot be withdrawn from, and if countries sign up to them they are subject to those treaties forever:

"1. A treaty which contains no provision regarding its termination and which does not provide for denunciation or withdrawal is not subject to denunciation or withdrawal unless:

- it is established that the parties intended to admit the possibility of denunciation or withdrawal; or
- a right of denunciation or withdrawal may be implied by the nature of the treaty.

2. A party shall give not less than twelve months' notice of its intention to denounce or withdraw from a treaty under paragraph 1."

It should be abundantly clear to you now that Treaties are 'international law' in action. A country surrenders a piece of its sovereignty when it signs and ratifies an international treaty, and no matter which political party you vote for, they cannot get that sovereignty back without a massive fight.

The US Constitution is effectively a Treaty between the individual states and also governs whether and how individual states can secede from the Union.[164]

---

163 The Vienna Convention on the Law of Treaties, 1969, http://legal.un.org/ilc/texts/instruments/english/conventions/1_1_1969.pdf

164 In a landmark 1868 ruling of the US Supreme Court in *Texas v White* (74 US 700), the court held that although the US Constitution prohibits unilateral secession, a state may leave the Union if the other states agree. Without that agreement, secession becomes rebellion and military force can be

In the new Totalitaria, this is precisely the framework that will allow the nice man from the United Nations to tie you up and blow raspberries at you. Your governments will plead powerlessness in the face of citizens protesting, because of their "treaty obligations". New Zealand, Britain, Australia and Canada have already surrendered their power, and if the US President and Congress do as well, there is nothing that can ever get that power back, short of a war.

The logic underpinning that statement is simple. Treaties impose obligations, backed up by the sanctions of international law. Those sanctions can include trade blockades, travel restrictions and even UN sanctioned military intervention. When the USA becomes just another subject state in a world federation, it too will be brought to heel by a global governing agency with the power to do to the US what the US used to do to other errant nations.

Again, the Greek and Cypriot bailouts give you a dry run demonstration. Athens fell apart, buildings burned, people were killed as angry citizens tore Greece to pieces. The financial and media commentators kept singing the same European song however – it's tough luck, if you don't agree to the conditions your economy will be left to die. Greece and Cyprus were left begging for the financial heroin they needed, and sure enough the fix was given.

Greece, after all, hasn't been powerful enough to win a war since Troy[165], so that option was pretty much closed off to them.

In Cyprus, the solution was particularly painful. The public were told that their bank accounts were being frozen, and each person with a savings or cheque account would be given a "haircut" – a euphemism for "we're raiding your bank accounts and taking some of your money". This was the edict from the European Union.[166]

The idea of staging a direct raid on the personal bank accounts of ordinary people has been toyed with before, but until now the technology has not been in place to make it possible. In fact, in New Zealand the Reserve Bank has outlined a similar plan if the banking system goes toes-up. It's called the Open Bank Resolution, and the theory is that if a bank finds itself in financial difficulties at 5pm, it is to alert the Reserve

---

used by the other states to quash it.
165 OK, that was a little unkind. Since Alexander the Great.
166 "There's Something Very Strange About The Cyprus Bank Haircut. Very Strange Indeed", by Tim Worstall, Forbes magazine, 31 March 2013, http://www.forbes.com/sites/timworstall/2013/03/31/theres-something-very-strange-about-the-cyprus-bank-haircut-very-strange-indeed/

Bank. The central bank will, in turn, freeze all accounts in the failing bank at that instant.[167]

Customers will not be told. Their online banking will suddenly stop working, as will phone banking, but the message will be given that the system is undergoing "maintenance" with apologies for the inconvenience.

During the night, Reserve Bank officials will liaise with the private bank executive team and assess how great the financial losses might be. They will then impose a "haircut" on the bank accounts of ordinary people of anywhere between 10% and 80% of the balance in the account. That money will be deducted to pay for the bank's accumulated losses.

At 9am the next morning the accounts will be unfrozen, with the deducted money already removed, and the public will be told for the first time that, while they were sleeping, XYZ Bank collapsed and required urgent re-capitalisation to preserve the integrity of the banking system.

Ordinary wage-earners, people supposed to pay rent and mortgages and grocery bills, will find that a big chunk of their funds are missing and that they will probably never see that money again. It would be particularly dark news if this event happened on the evening that your monthly salary went in to the bank account.

The Reserve Bank will pat itself on the back for quarantining that particular bank, but the flow on effects in the economy will be horrendous. Other companies will go unpaid because their customers suddenly cannot pay their bills.

As commentator Frank Macskasy noted, it's an unusual version of capitalism that shifts risk away from bank shareholders and on to depositors:[168]

"If this expropriation of deposits was ever to happen, do the depositors gain any benefit? Do they gain shares in the Bank as compensation? Or, if not, does that mean that shareholders gain the benefit of other people's money being used to prop up their investments?

"One could imagine an invalid on a WINZ benefit having his/her meagre savings 'taxed' to bail out a bank – to preserve an investor's shareholding that may be worth millions of dollars. This isn't justice or common sense, this is nasty, medieval, 'robber Baron' stuff.

"The biggest irony here is that, according to the principals of the free market, this is a kind of subsidy to a business – a subsidy enforced by the State, against the will of people who are not even shareholders in a particular bank.

---

167 http://www.rbnz.govt.nz/finstab/banking/4335146.pdf
168 http://thedailyblog.co.nz/2013/03/22/john-key-advocates-theft-by-banks/

"Even Marxists would balk at such extreme State power to seize people's money. They'd simply nationalise the bank and be done with it. Depositors would still have their modest savings left intact and untouched."

It turns out the measures the New Zealand Reserve Bank began planning for in 2011, and which rolled out in Cyprus in 2013, had their genesis in the G-20 summit meeting of major powers in South Korea, 2010.

Documents from that gathering of world leaders reveal an agenda item to plan for future financial collapse by shifting the debt burden in at-risk banks from taxpayers onto depositors. The G-20 nations have begun implementing this regime globally.[169]

The public fury at the events in Cyprus have given the politicians and bureaucrats pause for thought, but if the world goes into financial meltdown again because of the actions of Wall Street, the first people up against the wall will be you, the ordinary people who use banks on a daily basis.

As Bill Clinton's history professor at Georgetown University, Carroll Quigley, wrote in 1966, this has been on the agenda for a very long time:[170]

"The powers of financial capitalism had another far-reaching aim, nothing less than to create a world system of financial control in private hands able to dominate the political system of each country and the economy of the world as a whole. This system was to be controlled in a feudalist fashion by the central banks of the world acting in concert by secret agreements arrived at in frequent private meetings and conferences. The apex of the system was to be the Bank for International Settlements in Basle, Switzerland, a private bank owned and controlled by the world's central banks which were themselves private corporations."

The infrastructure has now been put in place globally. The legal mechanisms to legitimise raiding bank accounts have been passed. It could be the biggest transfer of wealth on a global scale ever seen, if it happens simultaneously. The shareholders of the global banks will have a large chunk of their debts paid by ordinary bank account holders, overnight. Let's face it, the United States teetered on the brink of financial oblivion in October 2013 and it owes trillions of dollars it simply cannot afford to repay. The Open Bank Resolutions being adopted worldwide are like the final stanzas in the tune for a game of musical chairs; you can sense the

169 http://www.globalresearch.ca/g20-governments-all-agreed-to-cyprus-style-theft-of-bank-deposits-in-2010/5335567
170 http://www.carrollquigley.net/pdf/Tragedy_and_Hope.pdf, page 324

music must be getting ready to stop, the fat lady must be getting ready to sing, but you don't know precisely when she will tip-toe onto the stage.

The mechanism of financial collapse may be deliberate, if Professor Quigley, who knew the players in the sixties and had access to some of their papers, is correct:

"This dominance of investment bankers was based on their control over the flows of credit and investment funds in their own countries and throughout the world. They could dominate the financial and industrial systems of their own countries by their influence over the flow of current funds though bank loans, the discount rate, and the re-discounting of commercial debts; they could dominate governments by their own control over current government loans and the play of the international exchanges," Quigley wrote. His description could equally have been applied to the 2008 GFC.

If the planet topples into financial turmoil again and depositors around the world find themselves fleeced, it would probably be an event sufficiently strong to usher in global governance and, in the ultimate irony, the public will probably beg for precisely that kind of rescue.

More to the point, the powers to do this are derived from international treaties, whose terms are then enacted in the signatory nations' legislatures. In 2009, however, the world came within a whisker of ushering in a global governing structure to end all global governing structures, and the mechanism being used to bring it was a climate change treaty.

Behold, over the page, the first UN treaty draft ever to document plans for a global "government".

CHAPTER 11

Climate Control – A Manufactured Crisis

"We are firmly convinced that the real danger lies not
in practical attempts, but in the theoretical elaboration
of communist ideas, for practical attempts, even mass
attempts, can be answered by cannon as soon as
they become dangerous, whereas ideas, which have
conquered our intellect and taken possession of our
minds...are demons which human beings can vanquish
only by submitting to them."
– Karl Marx[171][172]

In December 2009, the world's environmental and political glitterati gathered in the capital of the ancient kingdom of Denmark, for an event being billed as the last and only hope to avoid "an environmental calamity of Biblical proportions."[173] Another called it "the meeting that will determine the future of humanity".[174]

If it sounds religious in tone, it was. A pre-Copenhagen scientific conference of 2,500 delegates mysteriously alluded to an end-goal it called "the great transformation":

"Ultimately these human dimensions of climate change (the cultures

171 Communism and the 'Augsburger Allgemeine Zeitung' (Rheinische Zeitung, 15.10.1844.) in Collected Works (London. 1975) P. 221
172 http://www.net4dem.org/cyrev/archive/issue7/articles/Hudis/hudis.pdf
173 "Q&A: Copenhagen climate change conference 2009", The Telegraph, 19 November 2009
174 http://www.newscientist.com/special/copenhagen-climate-change-summit

and worldviews of individuals and communities) will determine whether humanity eventually achieves the great transformation that is in sight at the beginning of the 21st century."[175]

You will forget that phrase "the great transformation" as you read the next few pages dealing with the actual science of climate change, but when you get to the chapters on the motivating forces behind the push for global governance, you may well have a lightbulb moment. Climate change is not just a scientific concept, it is also a religious one, and a religion – ironically – that appeals to environmental scientists.

In that first week of December 2009, more than 15,000 officials, politicians and representatives from UN-affiliated Non-Governmental Organisations (NGOs), joined a 5,000 strong media contingent and a hundred world leaders. As the *Telegraph* newspaper reported, Copenhagen was also "blessed by the presence of Leonardo DiCaprio, Daryl Hannah, Helena Christensen, Archbishop Desmond Tutu and Prince Charles" to name a mere few.[176]

For Los Angeles-based diCaprio, it was possibly the first time he'd seen ice since filming *Titanic*; in true global warming fashion, the Danish capital was buried under nearly a metre of snow and temperatures were appropriately frigid. Nonetheless, Copenhagen was in party mood, and the climate delegates were getting ready to screw the world:

"Outraged by a council postcard urging delegates to 'be sustainable, don't buy sex', the local sex workers' union – they have unions here – has announced that all its 1,400 members will give free intercourse to anyone with a climate conference delegate's pass. The term 'carbon dating' just took on an entirely new meaning," reported journalist Andrew Gilligan.

To fuel the summit, organisers used CO_2, tonnes of the stuff. More than 1,200 limousines were on hand to drive delegates the few hundred metres from their hotels to the convention centre, and Copenhagen airport was home to more than 140 private jets belonging to people concerned about the climate but not concerned enough to use public transport to get there. A tally of the expected CO_2 emissions generated from the conference alone equated them to the total emissions of the English city

175 http://lyceum.anu.edu.au/wp-content/blogs/3/uploads//Synthesis%20Report%20Web.pdf
176 "Copenhagen climate summit: 1,200 limos, 140 private planes and caviar wedges" by Andrew Gilligan, The Telegraph, 5 December 2009, http://www.telegraph.co.uk/earth/copenhagen-climate-change-confe/6736517/Copenhagen-climate-summit-1200-limos-140-private-planes-and-caviar-wedges.html

of Middlesbrough and all its industry – that's a city of 143,000 people.[177]

The trappings of Copenhagen ultimately were not as important, however, as the issues on the table. At stake was a proposed international binding treaty on climate change, under the control of the UN Framework Convention on Climate Change (UNFCCC). This UN panel was set up in tandem with Agenda 21 at the 1992 Rio Earth Summit. It is the driving regulatory body set up to administer climate change rules for the UN.

America's National Public Radio, in a backgrounder to Copenhagen, set out some of the issues around the key areas of concern: who was going to administer the new climate regime, how would it be implemented, and who would pay?

"How much money will be available? Where will money come from? One idea is to make carbon a commodity that can be traded on international markets, and to tax each trade to build a fund for the developing world. That's highly contentious and may violate a U.S. prohibition against international taxes. Where will the money go? To the World Bank? The United Nations? Some new fund?

"If the nations of the world can agree on a framework for a climate, they'll need a set of rules to measure and define success," warned NPR.

Sure enough, there were firm plans being made to cover all those eventualities. In the third quarter of 2009, a draft of the proposed Copenhagen Climate Treaty was discovered in an obscure folder on the UNFCCC website. Clause 38 of that document caused eyebrows to raise worldwide – it specifically had the signatory States agreeing to set up a new world "government" to administer climate laws:[178]

The scheme for the new institutional arrangement under the Convention will be based on three basic pillars: government; facilitative mechanism; and financial mechanism, and the basic organization of which will include the following:

(a) The government will be ruled by the COP with the support of a new subsidiary body on adaptation, and of an Executive Board responsible for the management of the new funds and the related facilitative processes and bodies. The current Convention secretariat will operate as such, as appropriate.

(b) The Convention's financial mechanism will include a multilateral climate change fundincluding five windows: (a) an Adaptation window,

---

177 Ibid
178 http://unfccc.int/resource/docs/2009/awglca7/eng/inf02.pdf

(b) a Compensation window, to address loss and damage from climate change impacts, including insurance, rehabilitation and compensatory components, (c) a Technology window; (d) a Mitigation window; and (e) a REDD window, to support a multi-phases process for positive forest incentives relating to REDD actions.

(c) The Convention's facilitative mechanism will include: (a) work programmes for adaptation and mitigation; (b) a long-term REDD process; (c) a short-term technology action plan; (d) an expert group on adaptation established by the subsidiary body on adaptation, and expert groups on mitigation, technologies and on monitoring, reporting and verification; and (e) an international registry for the monitoring, reporting and verification of compliance of emission reduction commitments, and the transfer of technical and financial resources from developed countries to developing countries. The secretariat will provide technical and administrative support, including a new centre for information exchange.

It is the first time in history that a UN Treaty has purported to establish an international "government". This one was to be under the ultimate control of the COP, or "Conference of Parties", effectively meaning the signatories to the Treaty. Further on in the document, the new "government" was to be financed by a levy of 0.7% of GDP of developed countries. For American taxpayers, already up to their ears in debt, that means a further $105 billion in contributions to fund the new entity, but that's only the start. The Treaty proposed taxes on bank transactions and a range of other measures:[179]

(a) [Assessed contributions [of at least 0.7% of the annual GDP of developed country Parties] [from developed country Parties and other developed Parties included in Annex II to the Convention] [taking into account historical contribution to concentrations of greenhouse gases in the atmosphere];]

(b) [Auctioning of assigned amounts and/or emission allowances [from developed country Parties];]

(c) [Levies on CO_2 emissions [from Annex-I Parties [in a position to do so]];]

(d) [Taxes on carbon-intensive products and services from Annex I Parties;]

(e) [[Levies on] [Shares of proceeds from measures to limit or reduce emissions from] international [aviation] and maritime transport;]

(f) Shares of proceeds on the clean development mechanism (CDM), [exten-

---

179 Ibid, page 43. New Zealand taxpayers would be required to pledge a minimum of $1.5 billion a year in UN climate levies just based on the GDP levy alone, Australian taxpayers would be shelling out $10 billion a year.

sion of shares of proceeds to] joint implementation and emissions trading;
(g) [Levies on international transactions [among Annex I Parties];]
(h) [Fines for non-compliance [of Annex I Parties and] with commitments
of Annex I Parties and Parties with commitments inscribed in Annex B to
the Kyoto Protocol (Annex B Parties);]

The words in brackets were suggested additions or alternative wordings in the draft treaty document.

Remember, under the Vienna Convention on the Law of Treaties, if governments had signed off on this deal at Copenhagen and then ratified it in their parliaments, as is the usual process, it would have been binding and irreversible on all their citizens for all time. That's why it pays to watch treaty negotiations very, very closely.

The international transactions tax is known as the Tobin tax. The UN has been toying with getting this introduced since the 1990s; not surprising given it would generate around $1.5 trillion for a new global governance structure to use.

As you may have guessed, this treaty never got signed at Copenhagen in 2009. That's because British climate skeptic Lord Christopher Monckton found out about the draft treaty with its global government clause, and raised merry hell in the media. The Youtube video of his speech was viewed more than 2.7 million times[180] but, as the *Wall Street Journal* noted, "It deserves millions more because Lord Monckton warns that the aim of the Copenhagen draft treaty is to set up a transnational 'government' on a scale the world has never before seen.[181]

"The 'scheme for the new institutional arrangement under the Convention' that starts on page 18 contains the provision for a 'government'. The aim is to give a new as yet unnamed U.N. body the power to directly intervene in the financial, economic, tax and environmental affairs of all the nations that sign the Copenhagen treaty."

The revelations saw congress representatives and senators in the USA bombarded with emails and phone calls warning them of the electoral consequences of any sellout of US sovereignty to the United Nations.

President Obama evidently got the hint, because by the time he arrived at Copenhagen his speech specifically used the 'S' word to quell fears:

---

180 http://www.youtube.com/watch?v=PMe5dOgbu40
181 "Has Anyone Read the Copenhagen Agreement? U.N. plans for a new 'government' are scary", by Janet Albrechtsen, Wall St Journal, 28 October 2009, http://online.wsj.com/news/articles/SB1000 1424052748703574604574500580285679074

"We must have a mechanism to review whether we are keeping our commitments, and to exchange this information in a transparent manner. These measures need not be intrusive, or infringe upon sovereignty. They must, however, ensure that an accord is credible, and that we are living up to our obligations. For without such accountability, any agreement would be empty words on a page."

There was no agreement at Copenhagen in the end, but like rust the United Nations never sleeps. Unlike any other issue, the vexed debate about climate change and global warming has taken totalitarians closer to their goal than they've ever been. Climate change is a "big ticket" item, and by that I mean it's a subject that almost begs for a globalist solution.

The idea has permeated the media, political debate and the public consciousness that global problems need a global solution. What bigger problem could you confront the public with than "global warming"?

At the moment, there are two major initiatives underway. The first is supposedly a "scientific" one. This is the accumulation of research purporting to show that climate change is caused by humans, and that it is dangerous. Running alongside that is a political/policy narrative that aims to lock in governments, NGOs, big business and lobby groups into a 'framework' for combating the climate change peril.

The policy-wonks have, of course, already done their research on the supposed best way of tackling climate change politically, and that research has been presented almost as a fait accompli to politicians and regulators – either a carbon tax or an emissions trading scheme as a way of reducing CO_2 levels.

Skilled students will have recognised this as an example of the TINA doctrine in action: There Is No Alternative to a carbon tax/ETS, we must move now.

Whenever TINA is invoked, however, the response of smart people is usually, 'cui bono?', who benefits? Ducking down that rabbit hole you eventually come up against the totalitarians, of which more shortly.

First, however, let's review the state of the science on climate change, because that's what the whole debate ultimately hangs on. IF there's a crisis, and IF it's human-caused, and IF we can make a blind bit of difference to it, then there would be a legitimate argument to do so, regardless of totalitarian implications.

The question, obviously, hinges on 'IF'.

PROPOSITION ONE: EARTH IS WARMING UP

On the face of it, there is no doubt that the planet has been heating up over the past century or so. Historically, however, there is strong evidence this is mostly part of a natural cycle. The big scientific debate behind the scenes has not been about whether the earth is warming, but why?

There are two schools of thought on this, although only one gets media and political attention. The first view is that climate is mostly driven by natural cycles, many of which we do not fully understand yet. A thousand years ago, the world was hotter than it is today in what was known as the Medieval Warm Period.

About 800 years ago the world began to slip into what became known as "The Little Ice Age", where it was so cold that entire families would be frozen solid in their beds and even bottles of alcohol turned to ice. The Little Ice Age saw spectacular growth of glaciers worldwide, from Europe and the Americas to New Zealand and Antarctica. In Switzerland, entire villages were engulfed by walls of ice as the glaciers grew. Indeed, the glaciers we appreciate today hit their biggest size in millennia only a couple of hundred years ago, thanks to the cold temperatures. The mini ice age only ended around 1850.[182]

Both the Medieval Warm Period, and the Little Ice Age that followed were natural events. There were no factories or major human CO_2 emissions driving the climate during this period. Experts adhering to the "first view" believe the warming of the planet since 1850 is mostly a natural bounce-back from the viciously cold temperatures of the Little Ice Age, and only to be expected as the planet returns to a warmer climate. They point out that the glacier ice loss everyone is so worried about began happening in the 1800s as the climate warmed, and again is only to be expected. These glaciers are still much larger today than they were during the Medieval Warm Period, as climate scientist Doug Hoyt notes:

"As examples, the Aletsch and Grindelwald glaciers (Switzerland) were much smaller than today between 800 and 1000 AD. In 1588, the Grindelwald glacier broke through its end moraine and it is still larger than it was in 1588 and earlier years. In Iceland today, the outlet glaciers of Drangajökull and Vatnajökull are far advanced over what they were in the Middle Ages and farms remain buried beneath the ice."

---

182 Interestingly, Swiss glaciers have begun growing again, apparently because of the cooling of the first decade of this century and the heavy European snowfalls each winter. See *Switzerland on Sunday*, http://www.sonntagonline.ch/ressort/aktuell/3241/

That, then, is the first view in a nutshell: yes, CO_2 can have an impact but most of the warming we are seeing is natural, as the earth moves out of a six hundred year long cold cycle.

The second view, presented in the news media, and by the UN IPCC, and by Al Gore and other activists, is that CO_2 emissions are forcing global temperatures up.

As the argument goes, human population has ballooned out of control, we are raping our environment, polluting our surroundings, and pumping greenhouse gases into the atmosphere with every breath we take. As a result butterflies are migrating, polar bears are being trapped on tiny icebergs and all the warmth we are seeing is predominantly driven by human emissions which can only be stopped with a huge global bureaucracy enforcing new rules and the payment of new fees and taxes to ensure everyone does their bit.

Those taxes and fees will be paid to lobby groups like Greenpeace for further "research" and to so-called 'green-tech' companies whose investors include people like Al Gore. The climate-industrial complex and their political friends stand to make a lot of money and gain a lot of control.[183]

Totalitarian-spotters will be familiar with the narrative.

Supporters of this second view base their claims on a century old physics experiment where CO_2 in a closed system was found to absorb radiated heat and reflect it into the environment. Extrapolated out, the argument is that solar radiation hitting the earth's surface and being reflected back into space is being "trapped" before it leaves the atmosphere by CO_2 molecules in the air, creating a kind of blanket effect. In theory, there is nothing wrong with that argument.

In practice, however, the methodology is far more complex than this back-of-the-cereal-box simplistic story suggests. For a start, planet Earth is not enclosed, like a glasshouse. Heat can, and does, escape even after being captured by CO_2 gas molecules. Secondly, CO_2 is not the only ball in play when it comes to the climate.

We know this, because the latest figures show CO_2 levels in the atmosphere have continued to rise in a straight line from 380 parts per million 20 years ago to 400ppm today. According to orthodox global warming

183 There's huge money in the climate scare. The WWF raked in donations of US$2.6 billion over just five years between 2003-2007, by creating headlines and scare stories people could donate around. http://briefingroom.typepad.com/the_briefing_room/2010/02/look-how-much-money-green-groups-are-making-from-climate-scare.html

theory, such a rise will result in increased trapped heat. In fact, 350ppm was supposedly the trigger point for runaway global warming, which is why one climate change belief website calls itself 350.org.

However, a funny thing happened on the way to the oven: temperatures refused to rise.

Since 1998, world temperatures have flatlined. In fact, in recent years they've actually entered into a slight cooling trend. While every single child in the western world is exposed to global warming propaganda in their schools every single day, the incredible irony is that no one under the age of 15, anywhere in the world, has experienced global warming in their lifetime.

Let that sink in for a moment. No one under the age of 15 has seen global warming in action.

When Al Gore's movie *An Inconvenient Truth* filled cinemas in 2006/7, global warming was not actually happening. In fact, it was already beginning to cool.

When the news media trumpeted repeatedly "this is the hottest year ever", it turned out not to be the case. Global warming was not happening.

When politicians braved blizzard conditions and two metre snowdrifts in Copenhagen in 2009 to declare climate change as the defining issue of our time, global warming was not actually happening then either. Cynics would say, of course, that if Copenhagen delegates had simply looked out the windows they might have realised they were victims of cognitive dissonance.

The 2009 winter was so cold that old fashioned coal-burning steam trains had to be put back on Britain's main trunk lines because it was too cold for diesel and electric engines to run.

"When the temperature plummets and the snows start to fall which do you turn to – a traditional steam train or its multi-million pound modern replacement?," asked Britain's Telegraph.[184]

"Yesterday the steam locomotive, No: 45212, built in 1934, barrelled through the North Yorkshire countryside between Grosmont and Pickering, while hundreds of services on the country's modern electric network fell victim to the weather.

"Some train operating companies last night admitted that the computer

---

184 "Steam railways still running while modern network suffers weather disruption", The Telegraph, 4 Dec 2010, http://www.telegraph.co.uk/topics/weather/8181494/Steam-railways-still-running-while-modern-network-suffers-weather-disruption.html

software on their modern electric trains was not able to cope with the snowy and icy conditions."

Each northern hemisphere winter since 2009 has been bitterly cold, and researchers have been struggling to understand why temperatures have failed to rise in line with global warming predictions and continuing growth in CO_2 emissions.

Now, we have an explanation.

New research from respected climate scientist Mojib Latif and others shows the big warming periods like the late 1970s through the nineties, previously thought by climate scientists to have been caused by CO_2, were in fact most likely caused by natural cycles in the oceans, or what Latif calls "climate shifts".

This is particularly important, because the last IPCC report in 2007 said it could only detect a possible "human signature" in climate change since the 1970s. That claim was based on the assumption CO_2 was the primary driver. The latest research shows CO_2 had little if anything to do with warming since that time. Ergo, the "human signature" detected by the IPCC scientists does not appear to exist.

"These shifts…have a profound effect on the average global surface air temperature of the Earth," Latif says in a news release on his study. Changes in oceanic patterns turn "the world's climate topsy-turvy and are clearly reflected in the average temperature of the Earth."[185]

The UN IPCC's fifth assessment report, AR5, based its climate projections on computer models, and in particular a series of models known as CMIP5 which was described by its designers in 2012 as "a state-of-the- art multimodel dataset designed to advance our knowledge of climate variability and climate change. Researchers worldwide are analyzing the model output and will produce results likely to underlie the forthcoming Fifth Assessment Report by the Intergovernmental Panel on Climate Change."[186]

The most embarrassing aspect for the UN IPCC AR5 report is that CMIP5, the so-called "state of the art" simulation system anchoring the UN's climate projections, has failed epically to account for the massive

---

185 http://notrickszone.com/2013/08/26/leading-ipcc-scientist-concedes-oceans-have-profound-effect-on-average-global-surface-air-temperature/ Citing study "Hindcast of the 1976/77 and 1998/99 climate shifts in the Pacific" by Ding et al, *Journal of Climate* 2013 ; e-View doi: http://dx.doi.org/10.1175/JCLI-D-12-00626.1
186 "An Overview of CMIP5 and the Experiment Design". Bull. Amer. Meteor. Soc., 93, 485–498. April 2012
doi: http://dx.doi.org/10.1175/BAMS-D-11-00094.1

slowdown in warming over the past 15 years. In fact, the computer projections ran four times hotter for the period than the actual real observed temperature readings, as a just published report in the journal *Nature Climate Change* notes:

"The inconsistency between observed and simulated global warming is even more striking for temperature trends computed over the past fifteen years (1998–2012). For this period, the observed trend of 0.05 ± 0.08 °C per decade is more than four times smaller than the average simulated trend of 0.21 ± 0.03 °C per decade. The divergence between observed and CMIP5- simulated global warming begins in the early 1990s."[187]

The UN's AR5 report was out of date even before it hit the newsstands. AR5 claims a consensus higher than "95% certainty" that human-caused CO_2 emissions are predominantly driving global warming, but critics and even many scientists are now asking, "based on what evidence?" The assumptions the scientific "consensus" was supposedly built on are crumbling in the face of new evidence.

Confidence in the latest UN projections and journalistic fawning over them has not been enhanced by another new report in *Nature* suggesting the IPCC scientists are using statistical research techniques more than ten years out of date:

"Because the climate system is so complex, involving nonlinear coupling of the atmosphere and ocean, there will always be uncertainties in assessments and projections of climate change. This makes it hard to predict how the intensity of tropical cyclones will change as the climate warms, the rate of sea-level rise over the next century or the prevalence and severity of future droughts and floods, to give just a few well-known examples. Indeed, much of the disagreement about the policy implications of climate change revolves around a lack of certainty. The forthcoming Intergovernmental Panel on Climate Change (IPCC) Fifth Assessment Report (AR5) and the US National Climate Assessment Report will not adequately address this issue. Worse still, prevailing techniques for quantifying the uncertainties that are inherent in observed climate trends and projections of climate change are out of date by well over a decade. Modern statistical methods and models could improve this situation dramatically."[188]

187 "Overestimated global warming over the past 20 years", Fyfe et al, Nature Climate Change 3, 767–769 (2013) doi:10.1038/nclimate1972 Published online 28 August 2013
188 "Uncertainty analysis in climate change assessments", Katz et al, Nature Climate Change 3, 769–771 (2013) doi:10.1038/nclimate1980, Published online 28 August 2013

The so-called recently-discovered "climate shifts" work like this. Incoming solar radiation hits the 30% of the planet surface covered by land, and the 70% covered by water, and is either absorbed or reflected back into space to varying degrees. Much of the heat absorbed by the surface during the day is lost back out to space at night as the surface cools.

In the oceans, it's a little more complex. The heat that hits the ocean surface by day is absorbed by water, but water moves in currents, eddies and vortices. Some of that warm water is pushed by wave action and currents further into the depths, where the heat dissipates into the surrounding ocean, ever so slightly raising the temperature of surrounding waters. It can take a long time for what is left of that heat to be released back into the atmosphere.

Global warming believers like UN climate scientist Kevin Trenberth think the ocean is currently gobbling all the heat being thrown at it, and that "the heat will come back to haunt us sooner or later".[189] As others have pointed out, however, the issue is never that simple.

Trenberth's theory, which has been adopted by "believers", is that the heat is currently being stored in the deep oceans, at depths of 700 metres or greater. It has to be there, goes the argument, because there's no evidence of significant ocean warming in the layer between 700m and the surface – thousands of monitoring buoys known as the Argo project cover that sector.

So the argument as we know it comes down to this. Temperatures have failed to rise for the past 15 years and even UN IPCC scientists are acknowledging we have a "pause" in global warming. CO_2 emissions have risen strongly, but temperatures are actually falling at the moment. Kevin Trenberth and others argue that the missing heat must be going somewhere, and they theorise it is going into the deep oceans. It must be in the deep oceans, because it is not in the upper oceans close to the surface.

This argument, of course, begs another question. Precisely how did the supposed excess heat in the atmosphere over the past 15 years managed to transfer to the deep oceans without first raising the temperature of the shallows as it passed through?

The other twist to this climate fairy tale is the claim that the "heat" being stored by the oceans is, in fact, radiated heat from atmospheric CO_2. Ben Santer, another of the big names in UN climate science, made

189 "Missing" Heat May Affect Future Climate Change", Kevin Trenberth news release, 15 April 2010, https://www2.ucar.edu/atmosnews/news/2013/missing-heat-may-affect-future-climate-change

the bold claim in a 2006 study that most warmth at the sea surface was the result of greenhouse gases:

"Human-caused changes in greenhouse gases are the main driver of the 20th-century SST (sea surface temperature) increases," Santer's 2006 study claimed.[190]

Of course, this was music to the ears of those like Al Gore who wanted to attribute ocean-created events like Hurricane Katrina to human-caused global warming. It's a shame it wasn't true.[191]

Perhaps the easiest way of pointing out the error is an example of simple physics. The sun is the main source of heat on earth by a degree of considerable magnitude. Direct sun in the tropics can create surface temperatures hot enough to fry eggs on the pavement – 46 degrees Centigrade in the air and even hotter on dark asphalt. In contrast, reflected heat from CO_2 molecules (this is solar radiation that has already hit the earth and bounced back up into the atmosphere, so it's a fraction of the initial radiation, a mere 'heat shadow' if you like) is accused of causing global temperatures to rise around 0.8C over the past century and a bit, in contrast.

From this bare statement of fact, it follows as a point of logic that oceans will warm far more in response to direct sunlight, than they will from the miniscule blanket effect of greenhouse gases. Any study purporting to suggest that greenhouse gases are the "main driver" of ocean warming is therefore laughable.

One of the first to debunk it was Amato Evan. He and his team figured out, like you have, that the amount of heat that gets into the oceans is far more likely to depend on cloud cover and other variables, like dust, volcanic ash, pollution and smoke, that affect how much sunlight actually reaches the surface. Sure enough, when they plugged in the temperatures and atmospheric data for 26 years, they found ocean surface temperatures were far more influenced by these things than they were by greenhouse gases.[192]

"The tropical North Atlantic is unique among tropical ocean basins

190 "Forced and unforced ocean temperature changes in Atlantic and Pacific tropical cyclogenesis regions", Santer et al, September 12, 2006, doi: 10.1073/pnas.0602861103
PNAS September 19, 2006 vol. 103 no. 38 13905-13910
191 Although the UN IPCC 2007 AR4 report linked hurricanes to global warming, scientists are now rapidly back-pedalling, with "low confidence" that hurricanes and cyclones are a byproduct of warming. See "'Hurricane Marco Rubio' – A Winning Climate Campaign?" by Andrew Revkin, NY Times Dot Earth, http://dotearth.blogs.nytimes.com/2013/08/30/hurricane-marco-rubio-a-winning-climate-campaign/?smid=tw-share&_r=2
192 "African Dust over the Northern Tropical Atlantic: 1955–2008.", Evan et al, J. Appl. Meteor. Climatol., 49, 2213–2229. doi: http://dx.doi.org/10.1175/2010JAMC2485.1

because of its oftentimes extensive and heavy aerosol cover, a consequence of being downwind of West Africa, the world's largest dust source," wrote Evans.

In short, when there's plenty of dust being kicked up in Africa it keeps sea temperatures cool in the Atlantic, and when there's not much dust the sea temperatures rise – because of sunlight, not CO_2.

When they finally crunched the numbers, dust – or in fact the lack of it – accounted for about 69% of the warming in the Atlantic since 1980. Less dusty times meant more sunlight managed to hit the water.[193]

Armed with this knowledge of how the real world works, let's return to Mojib Latif's "climate shift" theory. Big hot and cold cycles within the oceans come around every so often and reset the climate system, effectively they are giant belches of heat into the air from huge areas of ocean, followed by periods of oceanic cooling.

As their names suggest, the Pacific Decadal Oscillation or Atlantic Multi-decadal Oscillation don't cycle heat in terms of months or weeks, but over decades.

"The AMO is an ongoing series of long-duration changes in the sea surface temperature of the North Atlantic Ocean," the National Oceanic and Atmospheric Administration (NOAA) website explains, "with cool and warm phases that may last for 20-40 years at a time and a difference of about 1°F between extremes. These changes are natural and have been occurring for at least the last 1,000 years.[194]

"Most of the Atlantic between the equator and Greenland changes in unison. Some areas of the North Pacific also seem to be affected."

Research has shown that in addition to those cycles, there are even longer term oscillations deep in the sea that can take centuries to circulate and release stored heat. Scientists studying ancient ice cores have found warming comes first as a result of solar cycles, and then about 800 years later CO_2 levels rise as the oceans get warm enough to release significant amounts of CO_2.[195]

---

193 If you are still wedded to the idea that a warming atmosphere caused by CO_2 is the most likely explanation for warmer seas, try this experiment at home: Position an illuminated 60w light bulb six inches above a glass of water for ten minutes. Measure the starting temperature and the finishing temperature. Then take another glass of water and breathe on the surface of the water for ten minutes (it's nowhere near exact, nor a direct comparison, but your warm breath will be significantly warmer than any CO_2 in the atmosphere would get in lab conditions in a controlled experiment using enclosed environments). Measure the temps.
194 http://www.aoml.noaa.gov/phod/amo_faq.php
195 "Deep-sea temperatures warmed by ~2°C between 19 and 17 thousand years before the

The importance of this cannot be overstated. It is peer-reviewed research from an IPCC-accredited research team, that essentially says the world's temperatures since the 1970s have been driven not by CO_2 at all, but by heat stored in the oceans. By definition, given the oscillation timescales, the heat emerging from the oceans in the 1970s must have been placed in the oceans decades, or even centuries earlier. Again, this means it cannot be related to human CO_2 emissions.

Whatever heat has emerged from the oceans to date has been driven by natural cycles, not man-made gases. And again, this means the much quoted IPCC claim that a "human signature" in climate change was detected after 1970 is no longer valid. It has been disproven.

In 2009, when my climate change book *Air Con* was published, I raised then what Latif and others have only now confirmed. *Air Con* cited a study by Alaskan researchers that made a similar point – Arctic melt appears to have almost nothing to do with CO_2 levels.

"Therefore, our conclusion at the present time is that much of the prominent continental arctic warming and cooling in Greenland during the last half of the last century is due to natural changes, perhaps to multi-decadal oscillations like Arctic Oscillation, the Pacific Decadal Oscillation, and the El Niño."[196]

The Pacific ocean El Nino of 97/98 created a massive hot spike in world temperatures. In fact, 1998 was listed as "the hottest year ever" by Al Gore. Leaving aside the fact that accurate temperature records are a very modern invention, thus making "hottest" a very dubious term, what actually followed the hot El Nino was a cold La Nina. It triggered what Mojib Latif calls a "climate shift" back into a cool phase.

"The most recent shift in the 1990s is one of the reasons that the Earth's

---

present (ky B.P.), leading the rise in atmospheric CO2 and tropical–surface-ocean warming by ~1000 years. The cause of this deglacial deep-water warming does not lie within the tropics, nor can its early onset between 19 and 17 ky B.P. be attributed to CO2 forcing. Increasing austral-spring insolation [higher seasonal solar radiation in the Southern hemisphere] combined with sea-ice albedo [heat reflectivity] feedbacks appear to be the key factors responsible for this warming." – SOURCE: "Southern Hemisphere and Deep-Sea Warming Led Deglacial Atmospheric CO2 Rise and Tropical Warming", L Stott, A Timmerman, R Thunell, Science 19 October 2007: Vol. 318. no. 5849, pp. 435 – 438 DOI: 10.1126/science.1143791

All very technical, but what this study found was that solar heat in the southern hemisphere warmed the oceans enough that ice melted and CO2 was released, but that it took up to a thousand years for the warmth to trigger CO2 release in any major way. In other words, far from CO2 being the "forcer" or instigator of warming, it was a result of warming that had begun a millennium earlier deep within the sea.

196 "Is the Earth still recovering from the "Little Ice Age"?", Syun-Ichi Akasofu, 7 May 2007, http://www.iarc.uaf.edu/highlights/2007/akasofu_3_07/Earth_recovering_from_LIA_R.pdf

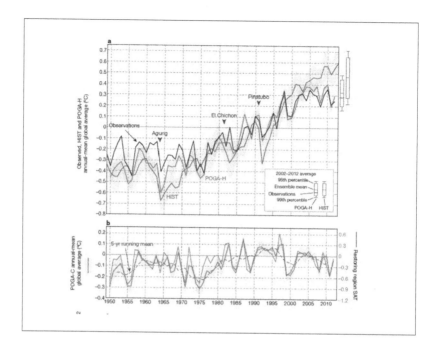

temperature has not risen further since 1998," noted the Latif study.

Roughly six months after I published *Air Con*, the Climategate scandal broke. One of the emails to emerge from that scandal in November 2009 and its follow-up leaks in 2011 and 2012 was this classic, particularly relevant to the revelations you've just read:

"What if climate change appears to be just mainly a multidecadal natural fluctuation? They'll kill us probably."[197]

And yet that's precisely the emerging scenario. In between typing the paragraphs above, and the sentence I am about to write, another major study was released, this time in the journal *Nature*. In a study headlined "Recent global warming hiatus tied to equatorial Pacific surface cooling", scientists Yu Kosaka and Shang-Ping Xie write:

"Despite the continued increase in atmospheric greenhouse gas concentrations, the annual-mean global temperature has not risen in the twenty-first century, challenging the prevailing view that anthropogenic forcing causes climate warming."

---

197 Tommy Wills, Swansea University to the mailing list for tree-ring data forum ITRDB, 28 Mar 2007 (email 1682)

To see a paragraph like that in a bastion of climate change propaganda like *Nature* is heartening of itself, but what emerged from their study was described by American climate scientist Judith Curry[198] as "mind blowing".

"Our results show," continues the study, "that the current hiatus is part of natural climate variability, tied specifically to a La-Nina-like decadal cooling." The study included a graph, tracking global temperatures against the oceanic phase they were investigating.

Judith Curry's commentary explains the significance. What you are looking for is the impact of "natural variability", as opposed to anthropogenic forcing (human-caused global warming):

"In Fig 1 a, you can see how well the POGA H global average surface temperature matches the observations particularly since about 1965 (note central Pacific Ocean temperatures have increasing and significant uncertainty prior to 1980).[199]

"What is mind blowing is Figure 1b, which gives the POGA C simulations (natural internal variability only). The main 'fingerprint' of AGW has been the detection of a separation between climate model runs with natural plus anthropogenic forcing, versus natural variability only. The detection of AGW [the 'human signature] has emerged sometime in the late 1970's, early 1980's.

"Compare the temperature increase between 1975-1998 (main warming period in the latter part of the 20th century) for both POGA H and POGA C:

- POGA H: 0.68C (natural plus anthropogenic)
- POGA C: 0.4C (natural internal variability only)

"I'm not sure how good my eyeball estimates are, and you can pick other start/end dates. But no matter what, I am coming up with natural internal variability associated accounting for significantly MORE than half of the observed warming.

"Like I said, my mind is blown. I have long argued that the pause was associated with the climate shift in the Pacific Ocean circulation, characterized by the change to the cool phase of the PDO. I have further argued that if this is the case, then the warming since 1976 was heavily juiced by

---

198 Judith Curry is the Professor and Chair of the School of Earth and Atmospheric Sciences at Georgia Institute of Technology. Curry is a respected climate scientist, but has earned extra respect from sceptics for not being "tribal" about climate science: she's prepared to follow the evidence where it leads, not just where she wants it to go. Her blog is essential reading for moderates in the climate debate.
199 http://judithcurry.com/2013/08/28/pause-tied-to-equatorial-pacific-surface-cooling/

the warm phase of the PDO. I didn't know how to quantify this, but I thought that it might account for at least half of the observed warming, and hence my questioning of the IPCC's highly confident attribution of 'most' to AGW," writes Judith Curry.

To put that in ordinary English, temperatures rose 0.68C during the study period overall, but 0.4C of that was entirely natural and resulted from the oceanic cycle – that's nearly 60% of the global warming being oceanic in origin, not CO2.

If you think that means the rest is caused by greenhouse gases, think again. The remaining percentage of warming can include other natural factors not tested for in this study.

More to the point, it backs up what Mojib Latif and his team discovered about oceanic "climate shifts" being the real drivers of atmospheric temperature.

Climate believers have been trying to spin these inconvenient studies for all they are worth, telling the news media and politicians that the oceans are eating heat and that's what's causing a temporary lull in global warming. What they are avoiding like the plague is the flipside of that same scenario: if natural oceanic climate shifts to a cool phase are the main trigger of global cooling, then natural climate shifts to a hot phase must equally be the main trigger of global warming, not humans.

But it's not just the ocean oscillations that have messed up the global warming story. Another big unknown is volcanism – both land-based and undersea. In fact, 95% of the world's active volcanoes are underwater. There are thousands and thousands of them bubbling and boiling away under the ocean, and 99% of the time we don't even know when they are erupting unless they happen to break through to the surface to create a new island.

Some of you may have noticed that in recent years there's been a big increase in seismic activity on earth – more earthquakes, more tsunamis and more volcanoes going off. Although we've always had them, scientists were startled to find a large number of unknown volcanoes exploding deep in the arctic ocean, creating volcanic glass along huge areas of the sea floor.

"Buried under thick ice and frigid water, volcanic explosions are shaking the Arctic Ocean floor at depths previously thought impossible, according to a new study," reported *National Geographic* in 2008.[200]

---

200 "Arctic volcanoes found active at unprecedented depths", by Kimberly Johnson, National

"Explosive volcanic eruptions were not thought to be possible at depths below the critical pressure for steam formation, or 2 miles (3,000 meters). The deposits, however, were found at seafloor depths greater than 2.5 miles (4 kilometers).

"Seismic activity was previously detected in the same region in 1999, along the Gakkel Ridge – a 1,200-mile-long (2,000-kilometer-long) mid-ocean mountain range north of Greenland. Hundreds of earthquakes were observed over a nine-month period, with magnitudes between 4 and 6. This earthquake swarm was the largest in recorded history along a spreading mid-ocean ridge and prompted researchers to return to the area for further investigation.

"In 2007 Sohn and his team stumbled across the glassy pyroclastic rock deposits while searching for hydrothermal vent fields in the Gakkel Ridge."

By definition, volcanic eruptions punching through the sea floor are capable of heating the deep ocean, although we don't know to what degree because we don't know enough about how active they all are. What's fascinating about the 2008 report is that it suggests these volcanoes are releasing massive amounts of CO_2 into the oceans:

"Ultraslow spreading ridge volcanic activity is infrequent, but the new findings of widespread rock debris suggest it could be exceptionally violent when it occurs, Sohn said.

"Such violence could be the result of a build-up of carbon dioxide underneath the ocean floor. If there was a bubble of carbon dioxide under the surface, the pressure would have built and eventually shattered through an earthquake-weakened crust, resulting in a volcanic eruption, Sohn explained.

"Sohn's team suggests that the amount of carbon dioxide would need to be at least ten times more than any other documented in seafloor samples in order to produce debris scattered over such a large area."

In 2012, scientists revealed they'd found massive mountain ranges and suspected volcanoes deep beneath the West Antarctic ice sheets, raising questions about whether the slide of glaciers to the sea in that area was really being triggered by reflected CO_2 warmth, or whether heat from the earth's mantle was in fact melting the ice sheets from underneath:

"It might have a big effect on the ice sheet and might explain some

Geographic, 26 June 2008, http://news.nationalgeographic.com/news/2008/06/080626-arctic-volcano.html

observations," seismologist Doug Wiens told journalists. "If you have a large heat flow from the mantle in a given area, it may form water at the bottom of the ice sheet."[201]

You hear all the time in the media and climate change reports how West Antarctic ice is at risk of melting. What they don't tell you as a rule is how much of a role increased volcanic activity under the ice is playing. The Pine Island Glacier, for example, is Antarctica's fastest moving glacier into the sea and has long been used as a symbol of human caused climate change, but again, it has a series of monster volcanoes boiling underneath it, only recently discovered.

A 2013 scientific study admits that geothermal heat appears to be a driver of ice melt under the West Antarctic Ice Sheet (WAIS):[202]

"The most common source of subglacial water is melting at the bottom of the ice sheet due to a combination of ice thickness, geothermal heat flux at the bed, and frictional heating due to rapid ice flow (Joughin et al., 2004; Llubes et al., 2006). Meltwater production could be enhanced by a higher geothermal gradient underneath parts of the WAIS (Shapiro and Ritzwoller, 2004)."

The significance of that is that volcanic activity can cause runaway melt because the water helps sluice the glacier out to sea much more rapidly, which is exactly what we are seeing.

"A volcanic eruption has the potential to produce large amounts of meltwater and, thus, could trigger a large flood event (Roberts, 2005; Bennett et al., 2009)."

A gravity survey of the earth's crust (the barrier of rock between us and the molten interior) underneath the West Antarctic Ice Sheet has found the crust has thinned dramatically under the ice where a continental rift appears to be opening up, which could be letting more volcanic heat into the ice sheet:[203]

"Major crustal thinning, coupled with low lithosphere rigidity, attest to the considerable impact of continental rifting beneath this part of the West Antarctic Ice Sheet...Narrow-mode rifting within the Pine Island

201 http://www.livescience.com/19546-antarctica-seismic-image-geology.html
202 "PaleoiceflowandsubglacialmeltwaterdynamicsinPineIslandBay,WestAntarctica" F.O.Nitscheetal, The Cryosphere, 7, 249–262, 2013 www.the-cryosphere.net/7/249/2013/ doi:10.5194/tc-7-249-2013
203 "Aerogravity evidence for major crustal thinning under the Pine Island Glacier region (West Antarctica)", Jordan et al, Geological Society of America Bulletin, December 30, 2009, doi: 10.1130/B26417.1
v. 122 no. 5-6 p. 714-726. http://gsabulletin.gsapubs.org/content/122/5-6/714.abstract

Rift is particularly important as it may serve as a geological template for enhanced glacial flow associated with Pine Island Glacier."

So far, researchers have established that the Pine Island volcano erupted massively two thousand years ago with enough force to punch entirely through the ice sheet and deposit ash fallout on an area of Antarctica larger than Wales. Significantly the volcano remains active, although we don't know whether it has been active all the time or whether it has only recently burst into life again.

One of the biggest clues that the volcano is deeply affecting the Pine Island Glacier is that the glacier is moving much faster than similar ice fields. Even so, media believers like the BBC insist on spinning Pine Island Glacier (PIG) as an icon of human-caused climate change:

"Satellite and airborne measurements have recorded a marked thinning and a surge in velocity of the PIG in recent decades. This has been attributed in part to warmer waters getting under, and melting, the ice shelf. The PIG's grounding line has pulled back further and further towards the land. The glacier's behaviour means it is now under close scrutiny, not least because it drains something like 10% of all the ice flowing off the west of the continent. 'The PIG is the most rapidly shrinking glacier on the planet,' explained Prof David Vaughan from the British Antarctic Survey (BAS). 'It's losing more ice than any other glacier on the planet, and it's contributing to sea level rise faster than any other glacier on the planet. That makes it worthy of study'."[204]

Carbon dioxide and ocean warming can take a thousand years to manifest; geothermal furnaces under the ice act much more rapidly.

A third factor, aside from ocean oscillations and volcanoes, is the sun itself. As mentioned earlier, it is the primary energy source for planet Earth. In 2009's *Air Con*, I pointed out that the sun has been more active in the past century than it has in the past one thousand years.[205]

"The Sun has been at its strongest over the past 60 years and may now be affecting global temperatures," reported solar scientist Sami Solanki. "The Sun is in a changed state. It is brighter than it was a few hundred years ago and this brightening started relatively recently – in the last 100 to 150 years.

"The unusually high number of sunspots during the past century sug-

204 "Antarctic's Pine Island glacier produces giant iceberg", BBC News, 10 July 2013, http://www.bbc.co.uk/news/science-environment-23249909
205 "Can solar variability explain global warming since 1970?", S Solanki and N Krivova, Journal of Geophysical Research, vol 108, 2003

gests that we currently may be seeing a state of the solar dynamo that is uncharacteristic of the Sun at middle age. Also, the higher activity level implies more coronal mass ejections and more solar energetic particles hitting the Earth. Thus we expect that the late 20th century has been particularly rich in phenomena like geomagnetic storms and aurorae."

When he and his team at the Max Planck Institute in Germany crunched the sunspot numbers they found the sun had become much more active after 1850, which coincided nicely with the end of the Little Ice Age and the beginning of the warm-up. Then towards 1900 sunspot numbers decreased again briefly and temperatures dropped. By the time the twentieth century was well underway, the sun was seriously powering up, and the planet's warming grew stronger with it, Solanki reported in another study.[206]

"Both [sunspot numbers] and temperature show a slow decreasing trend just prior to 1900, followed by a steep rise that is unprecedented during the last millennium; (2) great minima in the SN [sunspot number] data are accompanied by cool periods while the generally higher levels of solar activity between about 1100 and 1300 correspond to a relatively higher temperature (the medieval warm period)."

Again, in plain English, Solanki found that periods of minimal sunspots in recorded history coincided with cool climate like the Little Ice Age, while high sunspot activity coincides with warm spells.

These studies, published in 2003, were looking back at the twentieth century and finding "unprecedented" solar activity. Little wonder, you might say, that the planet has been warming up. The proof of any scientific theory is whether it works in all scenarios. Again, a funny thing happened on the way to the oven in the 21st century – the sun began powering down.

What was already apparent as a likely scenario when I wrote *Air Con* has today become undeniable:

"This year's [2013] solar maximum is shaping up to be the weakest in 100 years and the next one could be even more quiescent, scientists said Thursday (July 11). 'It's the smallest maximum we've seen in the Space Age,' David Hathaway of NASA's Marshall Space Flight Center in Huntsville, Ala., told reporters in a teleconference."[207]

---

206 "Millennium-Scale Sunspot Number Reconstruction: Evidence for an Unusually Active Sun since the 1940s", I Usoskin, S Solanki et al, Physical Review Letters, vol 91, no. 21, 21 November 2003, http://cc.oulu.fi/~usoskin/personal/Sola2-PRL_published.pdf
207 "Sun's 2013 Solar Activity Peak Is Weakest in 100 Years" by Megan Gannon, Space.com, 12 July 2013, http://www.space.com/21937-sun-solar-weather-peak-is-weak.html

There are many scientists who now believe it is not just the oceans eating global warming, but that in fact there is less warming fullstop because the sun is quieter now than it has been for a hundred years, and if the trend continues it will power itself down to Little Ice Age levels again. If that happens, say the experts, be very afraid; you will be praying *for* global warming, not against it. Remember, during the Little Ice Age American colonists were digging up graves and eating dead bodies to stay alive.

The Russians were the first in the world to pick that something was wrong. Way back in January 2008, while Al Gore was still basking in the glory of *An Inconvenient Truth* and climate belief lectures were playing to packed houses, Russian atmospheric scientists said their analysis of world temperature data revealed global warming had stopped, and the sun was becoming less active:[208]

"Temperatures on Earth have stabilized in the past decade, and the planet should brace itself for a new Ice Age rather than global warming, a Russian scientist said in an interview with *RIA Novosti* Tuesday.

" 'Russian and foreign research data confirm that global temperatures in 2007 were practically similar to those in 2006, and, in general, identical to 1998-2006 temperatures, which, basically, means that the Earth passed the peak of global warming in 1998-2005,' said Khabibullo Abdusamatov, head of a space research lab at the Pulkovo observatory in St. Petersburg.

"According to the scientist, the concentration of carbon dioxide in the Earth's atmosphere has risen more than 4% in the past decade, but global warming has practically stopped. It confirms the theory of "solar" impact on changes in the Earth's climate, because the amount of solar energy reaching the planet has drastically decreased during the same period, the scientist said."

At the time, their more famous climate colleagues in the West dismissed the Russian claims as nonsense. Every year was a "record" year, they insisted. "Global warming is happening, it's real, the debate is over," they insisted.

Of course, you've now seen that wasn't true, and the sharp-eyed amongst you will have seen from the dates of the scientific studies I've quoted that western climate scientists have only grudgingly come to that conclusion this year, not five years ago when the Russians did.

Let's see more of what scientist Khabibullo Abdusamatov had to say in 2008:

---

208 "Russian scientist says Earth could soon face new Ice Age", RIA Novosti News Agency, 22 January 2008, http://en.rian.ru/science/20080122/97519953.html

" 'A year ago, many meteorologists predicted that higher levels of carbon dioxide in the atmosphere would make the year 2007 the hottest in the last decade, but, fortunately, these predictions did not become reality,' Abdusamatov said.

"He also said that in 2008, global temperatures would drop slightly, rather than rise, due to unprecedentedly low solar radiation in the past 30 years, and would continue decreasing even if industrial emissions of carbon dioxide reach record levels.

"By 2041, solar activity will reach its minimum according to a 200-year cycle, and a deep cooling period will hit the Earth approximately in 2055-2060. It will last for about 45-65 years, the scientist added.

" 'By the mid-21st century the planet will face another Little Ice Age, similar to the Maunder Minimum, because the amount of solar radiation hitting the Earth has been constantly decreasing since the 1990s and will reach its minimum approximately in 2041,' he said.

"The Maunder Minimum occurred between 1645 and 1715, when only about 50 spots appeared on the Sun, as opposed to the typical 40,000-50,000 spots."

How is it that the Russians could accurately predict the flatlining of global warming and the powering down of the sun, when western scientists receiving some US$75 billion in funding failed to find this out until 2013? Was it incompetence, or wilful blindness caused by the glitter of all the funding gold being thrown their way?

In January 2013, US space agency NASA issued a grim warning:[209]

"Indeed, the sun could be on the threshold of a mini-Maunder event right now. Ongoing Solar Cycle 24 is the weakest in more than 50 years. Moreover, there is (controversial) evidence of a long-term weakening trend in the magnetic field strength of sunspots. Matt Penn and William Livingston of the National Solar Observatory predict that by the time Solar Cycle 25 arrives, magnetic fields on the sun will be so weak that few if any sunspots will be formed. Independent lines of research involving helioseismology and surface polar fields tend to support their conclusion."

If NASA, formerly a cheerleader of global warming, is now seriously suggesting that the sun is entering a phase reminiscent of the Little Ice Age, it's probably a good time to start investing in the wool industry.

As others have noted, the sun's activity in the current solar cycle is

---

209 "Solar variability and terrestrial climate", NASA, 8 January 2013, http://science.nasa.gov/science-news/science-at-nasa/2013/08jan_sunclimate/

currently at its "peak". That means what little solar activity still remains could be keeping global temperatures higher than they might otherwise be. As the peak dwindles back to low activity after 2014/2015, the real impact of a quiet sun on temperatures might begin to show much more dramatically.

A factor to add into this mix however, is time. We've already learnt that heat transferred into the oceans can take decades or even centuries to work itself back out into the atmosphere. Something that had long puzzled scientists, including some of those quoted above, was how a warming or cooling sun didn't appear to have a real-time connection to global temperatures. In a number of studies of warming post-1975, the scientists kept remarking, "Since 1975 global warming has occurred much faster than could be reasonably expected from the sun alone."[210]

Of course, you now know why: the warming between 1975 and the turn of the century was caused by oceanic climate shifts, which themselves had been driven by heat stored up from earlier times. The solar cycle was building up for the first half of the twentieth century and reached its most active around 1960. It now seems logical that some of that heat was stored in the oceans only to be regurgitated in the final quarter-century to create what we now call "global warming". It was a 'perfect storm' in the climate system – a stronger sun for the 20th century, pent-up heat in the deep oceans, and a tiny influence from humans on top.

We could not fully appreciate or understand the sun's role in that, until we understood the oceanic oscillation patterns better. And all this time, we thought it was CO_2 emissions driving temperatures up.

We've heard what the Russians were the first to accurately predict back in 2008, but have their views changed in any way? Apparently they've now identified three major cycles in solar activity – the 11 year solar sunspot cycles, a 90 year cycle, and a 200 year cycle:

"Evidently, solar activity is on the decrease," Yuri Nagovitsyn of the Pulkovo Observatory told the Voice of Russia radio programme.[211]

"The 11-year cycle doesn't bring about considerable climate change – only 1-2%. The impact of the 200-year cycle is greater – up to 50%. In this respect, we could be in for a cooling period that lasts 200-250 years.

210 "Phenomenological solar signature in 400 years of reconstructed Northern Hemisphere temperature record", Scafetta & West 2006, http://www.acrim.com/Reference%20Files/Sun%20 &%20Global%20Warming_GRL_2006.pdf
211 http://voiceofrussia.com/2013_04_22/Cooling-in-the-Arctic-what-to-expect/

The period of low solar activity could start in 2030-2040 but it won't be as pervasive as in the late 17th century".

If the Russians are correct, global warming will resume some time in the 2300s and will be a welcome relief for whatever is left of humanity at that stage. In the meantime, today's 15 year olds who have never actually experienced global warming will, by 2030, be the thirty-somethings tasked with guiding us through the darker, colder days to come.

WHAT ABOUT THE POLAR BEARS?

In August 2013, a picture of a dead, starved polar bear at Svalbard in northern Norway hit world headlines. It has been described, alternately, as "conclusive evidence of climate change", and "animal tragedy porn". A picture may well tell a thousand words, but the question one must always ask is whether the story being told is fiction or fact.

In the case of the emaciated bear, certain coincidences have emerged. Firstly, polar bear researchers had been in contact with that particular bear only three months earlier and reported it was "apparently healthy". To go from "apparently healthy" to literally a skin and bone carcass that "looked like a rug" in the space of 12 weeks, in an area of refrigerator temperatures and therefore very slow decomposition, seems to be stretching credulity much too far. Nonetheless, more coincidences emerge.

The dead bear just happened to have been photographed by a well-known polar bear climate activist who just happened to be in the right place at the right time. Cynics have suggested the photo was a PR stunt, jacked up by the researchers who initially stumbled across an ailing bear and waited for it to die.[212] The cynics may be right, or they may be wrong, it actually doesn't matter.

Despite the photo of the dead bear, polar bears by and large are thriving virtually everywhere in the Arctic, and this despite the best efforts of the World Wildlife Fund (WWF).

"Assuming the current rate of ice shrinkage and accompanying weight loss in the Hudson Bay region, female bears could become so thin by 2012 they may no longer be able to reproduce, " Lara Hansen, chief scientist for the World Wildlife Fund, told journalists back in 2007. This wasn't just any scientist, this was WWF's "chief" scientist.

Well, here we are, 2013, and the southernmost population of polar bears

---

212 Polar bears live to between 12 and 18 years in the wild. This one was 16. Old age is a probable cause.

in Canada, one living closest to the equator, is "doing just fine":

"Despite living under virtually identical conditions as the Western Hudson Bay subpopulation, touted by polar bear biologists and activists alike as the most severely affected by global warming, Southern Hudson Bay bears appear to be doing just fine," reports zoologist Dr Susan Crockford on the Polar Bear Science site. Interestingly, there was no significant difference in ice conditions between the Western Hudson and Southern Hudson regions.[213]

As for the Western Hudson Bay bears, an aerial survey has found they're not doing as badly as the climate scientists claim either:

"An aerial survey done in August by the Nunavut government, in response to pressure from Inuit, estimated the western Hudson Bay bear population at around 1,000," *The Star* in Canada reported. "That's about the same number of bears found in a more detailed study done in 2004. That study, which physically tagged the bears, predicted the number would decline to about 650 by 2011."

Commenting on the survey, Nunavut's director of wildlife management Drikus Gissing told journalists the climate change threat to bears was overrated:[214]

"People have tried to use the polar bear as a bit of a poster child – it's a beautiful animal and it grabs the attention of the public – to make people aware of the impact of climate change...We are not observing these impacts right at this moment in time. And it is not a crisis situation as a lot of people would like the world to believe it is."

That hasn't stopped a sizeable mythology building around polar bears and declining sea ice. Arctic sea ice has been melting more in summer, but that's been driven by two factors. One, warmer ocean temperatures for reasons already covered, and two, wind patterns blowing sea ice further south into the warmer currents. Regardless, the only season significantly affected is summer. During winter, spring and fall temperatures are so far below zero that sea ice is not an issue.

Does the vanishing summer sea ice cause problems for the bears? To an extent the answer has to be yes, but not as much as you probably think. The relevance of sea ice to polar bears is transport to and from feeding

213 http://polarbearscience.com/2013/07/25/southern-hudson-bay-subpopulation-status-farthest-south-of-all-polar-bears/

214 "Hudson Bay polar bear population defying predictions, Nunavut survey says", The Star, 4 April 2012, http://www.thestar.com/news/canada/2012/04/04/hudson_bay_polar_bear_population_defying_predictions_nunavut_survey_says.html

grounds. Analysis of polar bear populations has found most feed on ring seals primarily. When ring seal populations thrive, so do the polar bears that feed on them. When ring seal populations decline, so do the polar bears.

If you listen to the climate totalitarians, they'll tell you less sea ice is horrific for the bears. But truth be told, the times polar bears have done worst have been the years with the heaviest sea ice concentrations.

This may sound counter-intuitive, but follow the logic for a moment. It turns out that when sea ice is thicker and more extensive, little furry seal pups die because their noses can't find air:

"The seals must arrive before the new ice thickens in order to develop a series of breathing holes. When the ice first forms, the seals use their heads to punch open holes in the thin ice. Then as the fast-ice thickens, they must constantly chew and claw at the ice to maintain their breathing holes throughout the winter.

"Because seals require thinner ice to create their breathing holes, [215] areas dominated by thick multiyear ice always sustain far fewer seals and far fewer bears."[216]

By the time the global warming activists got hold of the story, it changed dramatically. A study published last year talked, under the heading "Effects of climate warming on polar bears", of their dietary dependence on ringed seal pups:

"In the mid-1970s and again in the mid-1980s, ringed seal pup productivity plummeted by 80% or more for 2–3 years…. A comparison of the age-specific weights of both male and female polar bears from 1971 to 1973 (productive seal years), to those from 1974 to 1975 (years of seal reproductive failure), demonstrated a significant decline in the latter period."

It's a sad and tragic story, until you realise what researchers Ian Stirling and Andrew Derocher failed to spell out: those years of seal pup decline and subsequent polar bear decline were years of heavy sea ice cover. In other words, the seals died because they couldn't break through the ice

---

215 Frost, K. et al. (2004) Factors Affecting the Observed Densities of Ringed Seals, Phoca hispida, in the Alaskan Beaufort Sea, 1996–99. Arctic, vo. 57. P. 115_128
216 Stirling, I. et al. (1999) Long-term Trends in the Population Ecology of Polar Bears in Western Hudson Bay in Relation to Climatic Change. Arctic vol . 52, p. 294-306.
See also Stirling, I. and Derocher, A. (1990) Factors Affecting the Evolution and Behavioral Ecology of the Modern. Bears: Their Biology and Management, Vol. 8, A Selection of Papers from the Eighth International Conference on Bear Research and Management, Victoria, British Columbia, Canada, February 1989 (1990), pp. 189-204.

to breathe, and the polar bears died because they couldn't break through the ice to catch seals.

To make it worse, the revelation that this is the fate of seals and polar bears in colder years came from one of Ian Stirling's own studies where he wrote: "the decline of ringed seal reproductive parameters and pup survival in the 1990s could have been triggered by unusually cold winters and heavy ice conditions that prevailed in Hudson Bay in the early 1990s, through nutritional stress".

Never let cold harsh facts get in the way of a cute bear story.

WHAT ABOUT RISING SEA LEVELS?

We often hear in the news about small Pacific Island nations struggling to keep their heads above seawater. As the story goes, sea levels are rising rapidly and the islanders are becoming the world's first "climate refugees". Proof of this is, ipso facto, proof of global warming theory, according to the story tellers on TV and in the newspapers.

Like all good yarns, however, there are two sides. Take climate poster-child Kiribati (pronounced Kiribarss) in the South Pacific. Photos are routinely published showing the ocean lapping at the very feet of palm trees. This is officially blamed on rising CO_2 causing catastrophic ice melt, in turn causing rising sea levels. As you've seen there is no catastrophic ice melt being caused by CO_2, and natural ocean warming cycles whose heat was absorbed decades to centuries ago are playing the dominant role. However, there's an even better explanation.

Coral atolls, which most of the South Sea islands are, form on the backs of living organisms – coral. The coral grows on the eroded rims of ancient volcanic seamounts, so the first and most important point to bear in mind is that most atolls are in tectonically active zones of the earth's crust. Takuu, north of Papua New Guinea, for example, although touted as another climate icon is in fact sinking rapidly because its volcanic seamount is slipping back under the continental plate (the immense weight of the seamount causes it to subside into the mantle).

The coral reefs form a protective ring around the atoll, a buffer against the big ocean rollers that would otherwise wash away the reef fish and the ecosystem inside the atoll lagoon. Once you damage the coral reef, however, you inevitably breach the system, just the same way as sandcastles on the beach dissolve faster if you break the protective wall around them.

Since World War 2, destructive techniques like dynamite fishing –

throwing explosives into the lagoon to stun fish – have taken hold, cracking the protective coral rings and allowing the oceans to begin rolling into the lagoons more frequently. In Kiribati in particular, dynamite fishing has been a huge environmental problem over the past 60 years and plays a large part in the coastal erosion it now faces.

There are other human-caused idiocies at play as well, however. In the Marshall Islands, for example, recently in the news, what the TV reporters didn't tell you is that large portions of the protective coral reef were harvested as "fill" for an airport extension. This is far from uncommon. Island populations have grown enormously since the war with the advances in trade and technology. Villages have become towns, and towns have become small cities. All that development requires resources and it has been cheaper, invariably, to mine their own coral reefs than to pay vast sums for imported material.[217]

The more coral removed from the reefs, the less protection the islands have from destructive ocean storms and large tides. The less protection they have, the more erosion they suffer, the more they need to harvest coral to replace what the sea keeps removing.

Then there's the problem of overfishing. With much larger island populations, but still an expectation of catching a feed in the lagoon, greater demand has been placed on the ecosystem. Studies have shown that parrotfish, for example, play a key role in creating coral sand by breaking down dead coral branches. That sand is then deposited on lagoon beaches by ocean activity, replenishing erosion. With a big drop in the number of fish, however, less coral sand is being created and deposited. When combined with the breach of the outer atoll, that means sand scoured from the beaches in storms can escape the lagoon entirely and never be replenished.

Vicious circle.

HOW DO YOU EXPLAIN THE RISING TEMPERATURES THEN?
Well, as already explained, satellite data is showing no global warming at all in nearly 17 years, but if you put that to one side you still have all these media reports talking about temperature records being broken. The reason for that is what's called the urban heat island effect.

---

217 "U.S. Taxpayers Funding Destruction of a Pacific Coral Reef", ENS, 22 June 2011 http://ens-newswire.com/2011/06/23/u-s-taxpayers-funding-destruction-of-a-pacific-coral-reef/

Most thermometers are located in towns and cities these days. Black asphalt, concrete and dark coloured roofs all absorb and store heat. At night, when the sun goes down, that built-up heat radiates quietly into the night air, making cities noticeably warmer, sometimes by up to six degrees centigrade, than surrounding natural countryside.

The thermometers used by weather bureaux and TV and radio stations are all in the cities, measuring this extra heat, and they're the reason it's officially 'warmer': in cities and towns, it actually is warmer, but it is not caused by CO_2. The data, however, is being used to support the CO_2 theory of global warming, which is utterly deceptive.

Climate scientist Dr Roy Spencer, who helps administer the satellite temperature research at the University of Alabama, Huntsville, has studied the temperature discrepancies caused by taking readings from within city boundaries:

"Based upon unadjusted International Surface Hourly (ISH) data archived at NCDC ... the warming trend over the Northern Hemisphere, where virtually all of the thermometer data exist, is a function of population density at the thermometer site.

"Depending upon how low in population density one extends the results, the level of spurious warming in the CRUTem3 dataset ranges from 14% to 30% when 3 population density classes are considered, and even 60% with 5 population classes."

In other words, up to 60% of the supposed increase in warming they talk about on TV is because the thermometers are situated in urban heat zones.

FINALLY, IF YOU'RE RIGHT WHY IS THE LATEST IPCC REPORT SAYING YOU ARE WRONG?

A funny thing happened on the way to the reality check – they left reality outside the door.

Leaked versions of the final draft of the IPCC's AR5 report had hit the news media late in 2013. Ironically, they hadn't been leaked to climate sceptics, but instead the IPCC itself had released a handful of copies to journalists it regarded as friendly, in the hope of generating positive spin about the climate change report.

Instead, one or two of the supposedly tame journalists reacted with horror when they realised the new IPCC report was admitting warming had stopped, and that there actually had been no warming at all this century.

They couldn't bring it upon themselves to dump on the IPCC directly, so the tame journalists slipped copies to sceptical journalists, who went on to have a field day with the draft AR5 report.

In damage control mode, the AR5 officials and political representatives gathered for a conference to "edit" the draft and remove the contentious bits about the 17 year pause, so that the final version would "stay on message".

As former New Zealand Minister of Science Barry Brill wrote in a commentary shortly after this fiasco, the justifications for editing the report were never going to stand up to scientific scrutiny:

"Under pressure at a media conference following release of its Summary for Policymakers, AR5 WG1 Co-Chair Thomas Stocker is reported to have said that *'climate trends should not be considered for periods less than 30 years'.*"[218]

This was a magic sleight of hand trick. Trends of two hot years in a row have been cited by climate scientists when it suited them. In fact, the year the IPCC was born there had been no global warming for the previous three decades, says Brill:

"When James Hansen launched the global warming scare in 1988, there had been no statistically significant warming over the previous 30 years and the warming trend during 1977-87 was 0.0°C. The IPCC was also established that year."

The IPCC's last report in 2007 used a 15 year trend, said Brill: "In 2007, the AR4 made much of the fact that the warming trend over the previous 15 years exceeded 0.2°C/decade."

In other words, the climate scientists are misleading the public now over what length of data is relevant, because an inconvenient truth has risen up to bite them.

Added to this, the AR5 report was already out of date on the morning of its release. Because of its editorial lead times, it did not include the latest studies set out at the start of this chapter revealing the impact of CO_2 had been vastly overestimated by the IPCC computer models.

Its data was wrong, and the AR5 report reflects that.

That, then, is the current status of the climate change debate. If you believe humans are the main drive of climate change, despite reading all the peer reviewed science just outlined, then you are a committed religious believer in climate change and its underlying philosophy.

So far in this book we've examined the 'how' of totalitarianism – how

218 http://wattsupwiththat.com/2013/09/30/to-the-ipcc-forget-about-30-years/

it is being implemented and the sort of laws now being enacted to enforce it. This next part of the book is where the story gets even more intriguing, even spine-chilling in places. You are about to find out why we're on the brink of a global totalitarian regime, what ancient human ambitions have driven it, and what its prime movers within the United Nations and various national governments really believe they are doing.

At the start of the book you were informed that the events detailed in this book actually happened. You need to hold onto that truth as you journey through this next section, because some of what you're about to read may shock you to your core. Pay careful attention to the footnotes and citations, they are there for a reason.

Many of you were not even born when some of these things were being devised in the fifties and sixties. Most of you will be unaware of all the events documented here.

There are elements in the chapters to come that *Da Vinci Code* author Dan Brown would have loved for his plot lines. The difference is, these elements are real. They are not fictional literary devices. These events took place, as described. This is not *The Omen* – a Hollywood distraction. This is a true story.

To apprehend the future, we have to understand the present, and to understand the present we must fully know the past. Who were we, what have we become, where are we going, who is leading us there? And so the next stage of our journey begins.

CHAPTER 12

The Earliest Control Freaks

"It's a dangerous business, Frodo, going out your door,' he used to say. 'You step into the Road, and if you don't keep your feet, there is no knowing where you might be swept off to'."

– J R R Tolkien, LOTR

With all great civilisations in human history, there are key moments which – when analysed through the rear-view mirror of scholarship – tell us how things were about to change. For the Aztecs, the arrival of Spanish explorer Hernando Cortes on the beaches of Mexico in 1519 was one such moment.

Cortes only had 600 men, 15 horses and 15 cannon, but he sealed his place in history by sinking his own ships in the bay off Velacruz. Why did he do this? Retreat was not an option. The event provides a case study in people management, and indeed is used in management training degrees and courses to this very day. It may have been 500 years ago, but the basics of human nature haven't changed.

By scuttling his ships before advancing, Cortes imposed his will on his men: they couldn't revolt because they were in a strange land, far from home, with no means of return transport. Many of them hated Cortes for that stunt, but their only option was to pledge loyalty and follow their leader wherever he may take them.

For the Aztecs, that event meant the end of their civilisation. Scholars today, favouring a politically-correct 'ethnocentric' worldview, attack Hernando Cortes for conquering the Aztecs and massacring thousands of them. On the other hand, a reality-check soon establishes that the Aztec empire was extremely bloodthirsty, having been built on human sacrifice, and that Cortes and his men were vastly outnumbered and needed – rightly or wrongly – to make bold military or political statements to keep the Aztec rulers on the back foot. This element of surprise does, of course, go back to the sinking-your-own-ships issue; it is a people management strategy still in use today and which you will discover as a permanent undercurrent as you navigate the pages of this book.

Returning to that issue of "key moments", however, while one such moment was the arrival of Cortes and his men, the real key moment was the sinking of the ships. That was the point that Cortes turned his expedition from a boy's own adventure into a life or death quest in full Darwinian style: kill or be killed, sacrifice or be sacrificed. From that moment, Cortes' men were shackled to him just as securely as if held by chains.

It was, as we say these days, the "tipping point". Readers familiar with Malcolm Gladwell's book of the same name will recall that huge changes often result from the activities of a relatively small number of people. Gladwell referred to the so-called 80/20 rule, in that 80% of the work is done by 20% of the people. It can of course be much lower. In the wider population, groups forming as little as two to five percent of the population can, when properly motivated, force change on the rest of society.

It's a little bit like inertia marketing – those annoying products you are finally persuaded to purchase late one night after watching an infomercial, complete with its "30 day money-back guarantee", and of course you never get around to sending it back. The same psychology applies to lobbyists and propaganda specialists. They know if they keep pushing, eventually the public will give in.

Most of us want peaceful lives; we are too busy to devote our full attention to whatever the *issue du jour* is in the media or on talk radio. We rely, often in blind faith, on the possibility that others with more time on their hands will get involved if there's anything requiring involvement.

Usually, change within a civilisation is incremental, not sudden. Evolutionary, not revolutionary. Ideas and civilisations die the death of a thousand cuts, not for them the glory of gladiatorial contest.

Of course, there are exceptions to this generality – the crushing of Nazi

Germany being one such – but as a rule the generality is generally true.

Which is what brings us to the precipice today: we are at the crossroads of modern civilisation. Everything humanity has ever done is a factor in that equation, a calculation towards the event that lies just around the corner. History is the sum of its parts, and right now it's throwing up some dangerous numbers.

For a start, we are living for the first time in a truly global environment. It has taken a while to get there.

The Babylonians were legends in their own lunchtime, and mostly in Babylon and Israel. The Romans may have been big, but they were only 'big' in Europe, the Mediterranean and the Middle East. The Spanish may have been big, but only really in the Americas and the Caribbean.

The British Empire, spreading out English-speaking settlers through Africa, Asia, the Americas and the Pacific, was a big step-up from the civilisations that came before, and set the scene for the twentieth century. At its peak in the early 1920s, nearly a quarter of the globe was under the control and authority of the King of England, and more than 20% of the total world population. While aging colonels sipped Singapore Slings at Raffles Hotel and reminisced about campaigns against "the Fuzzy-Wuzzy"[219], the British civil service grew at a dramatic rate to administer an empire "on which the sun never sets" – a reference to that fact that at any given moment, somewhere in the world, the sun was shining on a British subject.

To put the British empire in perspective, the Romans covered only 4.3% of the world's land area –pretty much the same size as that controlled by Hitler's Germany. At its peak, Spain accounted for 13% of the globe. Britain, just to remind you, had 24% of the whole planet.

The only empire to come close – very close in fact – to the British empire is the Mongols. Their power stretched from China across Russia and into southern Asia and eastern Europe. At one stage in the mid 1200s the Mongols overran Baghdad and toppled the Islamic empire of the day, the Abbasids.

The push west by the Mongol's ruling Khan dynasty was relentless. Their armies surged through Bulgaria and Hungary almost to the gates of Vienna in Austria before being turned back by Christian forces. In

---

219 A late nineteenth century slang term for the Hadendoa tribe of East Africa, so named for their hairstyles. The tribe contributed to a Muslim uprising in the Sudan, and its men were feared fighters who routinely mutilated the bodies of dead British soldiers.

fact, Christian and Muslim armies engaged in the Crusades temporarily buried their own rivalries in the Middle East and joined forces to defeat the Mongols at the battle of Ain Jalut, near Galilee, northern Israel, in 1260AD.

Had the Mongols succeeded in vanquishing Europe in the thirteenth century, history might have taken a very different turn. Instead, as already noted earlier, the story of the West is built on tipping points like these. When its fate hung in the balance, the Mongol empire crumbled. To the victors the spoils, and with it the rights to tell the stories.

At their peak, the Mongols controlled almost as much of the globe as the British later did; yet today, you'd barely know their civilisation ever existed. Empires rose and fell over the millennia of human civilisational development, but something always prevented them from reaching true critical mass. Competition. As long as there was competition, no one civilisation could be all-powerful all the time.

There were other mitigating factors at play, as well, however.

It wasn't until the advent of reliable shipping that any genuine prospect of real world domination existed. While the Vikings had managed to get to Greenland and North America, their longboats were not the most seaworthy craft in North Atlantic storms.

The Spanish and Portuguese changed all that. It wasn't until the 1400s that Europeans managed to design genuine ocean-going ships. For centuries they'd been limited to coast-hugging vessels that needed the shelter of a nearby harbour at all times. They might have been suitable to cross the relatively benevolent Mediterranean or scoot up the English channel, but not to explore the great unknown beyond the strait of Gibraltar. [220]

The Mongols had known nothing of boats in any meaningful sense.

---

220 On the other side of the world (in fact directly opposite Spain if you drilled straight through the earth), about 200 years earlier, Maori tribes from Polynesia migrated to New Zealand, the last major landmass in the world to be settled by humans. Their voyages were made on giant sail-powered catamarans. Captain Cook reported seeing such double-hulled canoes in both Polynesia and Fiji. Some, up to 36 metres in length, were actually bigger than Cook's bark *Endeavour* at 33 metres, and considerably faster in the right conditions. There is evidence of two-way traffic between New Zealand and Polynesia, across some of the roughest cyclone-infested ocean in the world. To put the Polynesian navigation into perspective, the distance from Tahiti (thought to be the ancestral Maori homeland 'Hawaiiki') to New Zealand is 4,100 km, whereas the distance from Ireland to the North American coast is only a little over 3,100 km. Columbus' journey from Spain to the Bahamas clocked out around 6,000 km. Modern catamarans, like those used in the recent America's Cup campaign, owe the genesis of their design to those ancient Polynesian craft. Again, to put them into size perspective, the America's Cup cats were 72 feet (22m), whereas the Polynesian cats were 118 feet (36m) in length.

Their empire ended on the sand of China beach at the Sea of Japan; horses were the ships of the steppes carrying the Mongol imperial forces to their destinations far away.

The Romans, likewise, had been largely a terrestrial rather than maritime power. The Roman galleys were a step down the evolutionary ladder from the Viking longboats and open sea voyages were even more a game of Russian roulette. For a thousand years, then, civilisations had been built by foot or hoof, and that limited what a civilisation could achieve in the great scheme of things.

By the mid 15th century, however, innovative tweaks in boat design had resulted in two major improvements, the caravel and the carrack. These are the kinds of sailing ship that opened up exploration of the new world in the same fashion that the Saturn V rocket opened up exploration of the Moon five hundred years later.

When Christopher Columbus discovered the West Indies in 1492, his fleet was one carrack (the larger *Santa Maria*) and two caravels, *Pinta* and *Nina*. When he set foot in what we now call the Bahamas for the very first time, it truly was one small step for a man, but a giant leap for mankind. It's all very well to complain about Eurocentrism – the belief that the world revolves around Paris or London –but the first civilisation in modern history to make a lasting, there-and-back-again impact on a continent over the ocean was European.[221]

The possibility of a global civilisation was edging nearer, but it wasn't there yet.

---

221 British author Gavin Menzies has theorised in his book "1421: The Year China Discovered The World" about a great Chinese 'fleet' that was sent out to explore the world in 1421 and discovered the US, New Zealand and Australia. Menzies' lack of fluency in the Chinese language, coupled with a disturbing lack of hard evidence to support the main contentions, led historians to reject the theories as unproven and even "fictitious". Similarly, other authors theorising Phoenician or ancient Libyan voyages that supposedly discovered Australia and New Zealand also run into credibility difficulties: their boats simply were not advanced enough. It was possible for a Phoenician ship to explore by hugging shorelines, and technically that could get you all the way from the Middle East to Australia via the coast of Asia, but it could not get you across the open ocean of the 2,200 km Tasman Sea that lies between Australia and New Zealand, let alone get you back again. Of course, even small yachts make those kinds of voyages now, but there are light years of development between a 2,500 year old open longboat and an enclosed yacht. The Phoenicians circumnavigated Africa, their voyage described in the writings of Herodotus. However, as the website Phoenicia.org notes, "We have, however, no direct evidence that their commerce in the Indian Ocean ever took them further than the Arabian coast." http://phoenicia.org/ships.html

CHAPTER 13

On Pain Of Death

"Any man or woman who robs any garden, public or
private, while weeding it or who wilfully pulls up any
root vegetable, herb, or flower to spoil or waste or steal
it, or robs any vineyard or gathers the grapes, or steals
any ears of corn ... shall be punished with death."
– Sir Thomas Dale, Virginia Governor, 1614

With the Age of Discovery came the age of conquest. The Spanish rapidly spread through Central America, Mexico and eventually into Texas, California and other southern regions of the US. The British, meanwhile, were colonising the north, or trying to.

In 1584, only 92 years after Columbus opened up the new world, Britain's Queen Elizabeth 1 and her adviser Sir Walter Raleigh sent colonists to establish a town in the Americas. Partly the idea was altruistic in terms of settlement, but also devious, in terms of providing a base from which to raid Spanish ships transporting gold back home from the new world. Britain knew that the vast wealth now at Spain's fingertips could play a massive role in tipping the global balance of power, and they couldn't afford to be passive while Spain engorged on gold and gems. In those days, nations lured peasants with promises of fat purses in the agricultural off-season to join the army or navy for a time. Mexican gold bought a lot of mercenaries, and built a lot of warships.

Civilisation has always been a manifestation of power.

The English colonists' fleet crossed paths with the Spanish at Puerto Rico before heading north to Roanoke Island off the coast of what is now North Carolina.

It was an ill-fated, star-crossed attempt, not least of all because it coincided with the outbreak of the Anglo-Spanish War. The Spanish, sensing that a race for global power and influence was on, had raided Sir Francis Drake's slave shipments a decade or so earlier, and in retaliation England had stepped up what they called "privateering" – effectively piracy by officially-sanctioned private individuals and companies. It was the forerunner of what we might call a hostile corporate raid. Privateers were pirates who attacked the enemy's ships, rather than their own, receiving an official blessing in return for sharing some of the spoils with the Crown.

Added to the already volatile mix was a dispute between England and Spain breaking out on religious lines. England's decision under Queen Elizabeth's father, Henry VIII, to set up the Anglican Church in defiance of the Catholic Pope infuriated the Spanish and the French monarchs – both staunch supporters of Rome.

Spain's King Philip sent agents to begin stirring up a Catholic rebellion in English-controlled Ireland, and when England retaliated by agreeing to aid Protestant rebels in the Spanish-controlled Netherlands, more fuel was added to the fire. Spain saw it as an outright declaration of war. In 1585, the same period as the Roanake Island colony was being established, Sir Francis Drake led attacks on Spanish towns and ports in the West Indies and even hitting the St Augustine settlement in what is now Florida. The cold war had become a hot war.

Every British warship available was recalled to defend the homeland when the Spanish retaliated with their Armada in 1587. By the time the dust settled and peace was finally declared in 1604, the war had raged through Spain, England, Ireland, France, the Netherlands and Belgium, Portugal, the Americas, the North Atlantic and even the Canary Islands. It was a battle for power, for money, for belief.

And by the time the ink was dry on the peace treaties, the forgotten colonists left behind at Roanoke Island had vanished, becoming forever known as "the lost colony". They earned their place in history, however: tobacco, sweetcorn and potatoes[222] were introduced to England from the

---

222 In the civilisational vegetable arms race, Spain can lay claim to the first commercial use of potato crops in 1570, from tubers brought back from its own territories in the Americas

Roanoke farms before the war made ongoing contact impossible.

What happened to the missing colonists?

Resupply ships sent from England had been captured by the Spanish. The first 107 colonists – all men – had hitched a ride home with Sir Francis Drake in 1586, bringing with them the aforementioned potatoes. When one ship got through to Roanoke in 1587, they dropped off a hundred and fifty more colonists, including newborn baby girl Virginia Dare, the first English child born in America. Virginia's grandfather, Governor John White, returned by ship to England to urgently argue the case for reinforcements and supplies for the struggling colony.

Again, because of the war, he was delayed, and did not set foot at Roanoke again until 18 August 1590 – what should have been Virginia's third birthday. Only, she wasn't there. Nor were any of the other men, women and children who'd been left behind at Roanoke. Nothing had been burned, there was so sign of violence, but the entire community was gone, never to be seen again.

Regardless of their loss, the British were not about to give up on expanding their civilisation into the Americas. The recent peace treaty with Spain gave breathing space for the Virginia Company of England to despatch a fresh team of colonists. This time they landed in a swamp on the shores of Chesapeake Bay in Virginia in 1607, and they named their new colony Jamestown.

Once again, it wasn't going to be easy. Plagued by malaria, poor agricultural conditions and a group of English settlers unused to hard manual labour, death soon stalked the colony. More than fifty died in the first few months of arrival. It was the height of the Little Ice Age – a global freeze-up that made life harsh even at the best of times. For Jamestown residents, these were not the best of times, not even close. The winter of 1609-1610 was so bleak that many starved, and those who didn't resorted to cannibalism[223], as a journal article in the Virginia archives describes:

"...driven through insufferable hunger to eat those things which nature most abhorred, the flesh and excrements of man as well of our own nation as of an Indian, digged by some out of his grave after he had laid buried there days and wholly devoured him; others, envying the better state of body of any whom hunger has not yet so much wasted as their own, lay wait and threatened to kill and eat them; one among them slew his wife

---

223 "Skull proves settlers resorted to cannibalism", ABC Australia, 2 May 2013, http://www.abc.net.au/news/2013-05-02/cannibalism/4664156

as she slept in his bosom, cut her in pieces, salted her and fed upon her till he had clean devoured all parts saving her head…"[224]

Of the 500 or so Jamestown settlers in 1607, by 1610 only sixty were still alive. But still they didn't give up.

In a similar display of human spirit to that demonstrated by Hernando Cortes when he sank the ships that could have taken them home, the Jamestown colonists pushed on, heavily reliant on resupply ships from England. A foothold in the new world, however tenuous, had been established.

Although the Jamestown site was abandoned in 1699 in favour of nearby Williamsburg, it served as capital of the American territories up until that year.

Migration to America was opening up rapidly, despite the known hardships of famine, disease, bitter cold and skirmishes with Native American tribes. In 1620 the pilgrim ship the Mayflower landed at Cape Cod in New England carrying 150 passengers and crew – the much documented "Pilgrims" voyage. By the time they settled at New Plymouth some weeks later, around half had died.

A second voyage of pilgrims in 1629 disappeared without trace, taking 140 men, women and children with it.

It seems strange in this age of jet transport, homeland security checks and proposed space planes, that travel used to be an utter crap shoot. Nonetheless, that's exactly what it was and people were still strong enough to chance it.

Take, for example, conditions on a 1750 voyage from Europe to America:

"During the journey, the ship is full of pitiful signs of distress – smells, fumes, whores, vomits, various kinds of seasickness, fever, dysentery, headaches, heat, constipation, boils, scurvy, cancer, mouth-rot, and similar afflictions, all of them caused by the age and the high salted state of the food, especially of the meat, as well as by the very bad and filthy water… Add to all that shortage of food, hunger, thirst, frost, heat, dampness, fear, misery, vexation, and lamentation as well as other troubles…On board our ship on a day in which we had a great storm, a woman about to give birth and unable to deliver under the circumstances was pushed through one of the portholes into the sea."[225]

224 Journals of the House of Burgesses, Virginia, republished by Howard Zinn in "A People's History of the United States", 1980, see http://libcom.org/history/1619-1741-slavery-slave-rebellion-us
225 Zinn, "A People's History of the United States", HarperCollins edition, 2005, page 43, citing German migrant Gottlieb Mittelberger

Of course – and this is highly significant – such pioneering voyages in the 1600s took place in an age where mass media did not exist. If your neighbour, or even your children, gained passage to the Americas it was usually a one-way trip; you might never hear from them again. No telephones existed, the telegraph had not been invented, even an official postal service would not arrive on the scene until 1840. Much of the populace was illiterate, and radio, TV, and newspapers[226] as we understand them today did not exist either.

As a result, pioneers often took these voyages in the absence of a full appreciation of the risk. No cameras were on hand to broadcast footage to a waiting world of a distressed pregnant woman being thrust through a porthole to her death in the cold Atlantic ocean; if it were not for the writings of passenger Gottlieb Mittelberger we would never have known.

If your kin went to America or the other colonies, chances are you would not find out what became of them. Did their ship sink without trace on the voyage? Have they gone and built a log cabin in the woods or the prairies somewhere? People had questions but often no answers, and certainly nothing more than generalised rumours to put them off making the journey. In the absence of wall-to-wall media coverage of every scary event, the public's willingness to step into the great unknown, often on faith, was far higher than it is today. And even if they knew the risks, were their present squalid circumstances anything to boast about if they stayed in England or Europe?

Think on that point for a moment. The defining events of human history have mostly been achieved without the scrutiny of media coverage or direct government oversight. People on the ground at a special moment in time saw an opportunity, weighed up the risk as they understood it, and seized the day.

The questions you need to ask yourself are these: could and would today's generation take such risks, or have we become civilisationally gunshy? Are we now too reliant on Nanny State to protect us from risk? For the first time in human history, have we lost our mojo? Or have soft furnishings, wall-to-wall entertainment and flushing toilets made us just a little bit too comfortable?

We've seen how our forefathers, in all cultures and races, were not

---

226 Small presses printed dozens of small papers in London, often only four pages in size and mostly advertisements, but these relied on rumour and gossip, not first hand accounts from overseas news bureaux

paralysed by fear. Life was short and often brutal. You made of it what you could.

There is a paradox in all this, however, and it is this: colonies like America and Australia were built on political and administration systems that were utterly undemocratic and effectively totalitarian in nature. Feudal lords governed the US colonies, and the penalty for misdemeanours was often hanging. Arguably, it was no worse than back in mother England where Parliament since the late 1600s had increasingly chosen to lay out in statute what had until then been "common law" offences. Suddenly, stealing anything worth more than tenpence was an invitation to swing from a rope.

This was a transitional time in England. Although the state clearly existed, as did the King and the King's men, for the most part ordinary commoners were largely untouched by Government as no income taxes operated and there was only a tiny bureaucracy, largely confined to collecting Customs duties.

Even so, where harsh state decrees ordered the death penalty for minor offences, juries and courts often took sympathy on the perpetrators. In his essay "The Georgian Underworld", UK writer Rictor Norton suggests juries of ordinary people provided a useful counterbalance against what could be seen as authoritarian rule:[227]

"Capital convictions and executions steadily declined over the course of the eighteenth century. Although the code, potentially, was sanguinary enough to strike terror into the hearts of potential criminals, in actual practice it was largely merciful and lenient. The net effect of explicitly making more crimes punishable by death was to focus the minds of jurors upon the seriousness of their deliberations and the importance of proving crimes beyond the shadow of doubt. Even prosecutors and witnesses were less willing to press charges or testify when they realized they would be sending someone to the gallows for crimes that threatened only property rather than life.

"Both judge and juries were sometimes distressed that theft of a very small amount was a capital offence (i.e. requiring the death sentence), and they often either withheld a guilty verdict even when guilt was clearly established or arbitrarily withheld the death sentence. A not-untypical case involved Ann Flynn who in 1750 was indicted for stealing a shoulder

---

227 See http://rictornorton.co.uk/gu02.htm

of mutton from a butcher in Whitechapel," writes Norton.

The contemporary record of Ann Flynn's dice with death is found in *The Newgate Calendar* – a compendium of crimes and morality stories from the 1700s:

ANN FLYNN was indicted at the Old Bailey for stealing from a butcher in Whitechapel a shoulder of mutton. It appeared in evidence that, the prosecutor being busy with his customers on a Saturday night, the prisoner availed herself of that opportunity, and carried away the shoulder of mutton. She was, however, soon seized and brought back, and, an officer being sent for, she was carried before a magistrate, and committed for trial.

These facts being proved, the prisoner was called upon for her defence; and she told a tale of woe that penetrated every heart. She acknowledged the robbery; but solemnly declared she was urged to it by the most afflicting distress. Her husband had been ill and unable to earn a shilling for twelve weeks, and she was driven to the last extremity, with two infant children. In that deplorable situation, continued the unfortunate woman, while the tears ran down her wan cheeks, she desperately snatched the shoulder of mutton – for which she had already been confined five weeks.

The jury found her guilty, with a faltering accent; and the recorder immediately replied, "Gentlemen, I understand you," and sentenced her to be fined only one shilling and discharged, which the jury themselves paid, but the officer of the prison gave it to her.

This case, if the extremity of the law had been resorted to, was felony.

As soon as she was taken away, the prosecutor addressed the Court, and said that the constable had done him more injury than the thief; for though Sir William Parsons, the magistrate that committed her, had ordered him to take care of the shoulder of mutton, he thought fit to cook it for his own dinner, and to sit down and eat it.

[This new complaint, as might naturally be supposed, excited not a little the risible muscles of the Court.]

The constable was immediately called upon to account for his conduct, who said: "My Lord, I did take care of it, as ordered; I kept it whilst it was worth keeping, and if my wife and I had not eaten it, the dogs must have dined on it."[228]

Juries and courts, regarding the penalties as often too harsh, frequently

---

228 From the book "The Newgate Calendar", published in the 1700s. See http://www.exclassics. com/newgate/ng231.htm

found as a 'fact' that the value of the property stolen was less than 10d., even when it was patently worth more, thereby taking hanging off the table as a punishment. Additionally, many of those sentenced to death for various crimes also had those sentences commuted, provided they agreed to be transported to the colonies in America and later Australia.

Britain had tossed up whether to establish its replacement penal colony, after losing the US war of independence, in New Zealand, but determined the native Maori inhabitants were far too numerous, far too intelligent and far too dangerous – having seen off the French navy in one confrontation. The safety of a penal colony in New Zealand could not be assured, officials decided. Australia became the new convict colony in 1788.[229]

These were tough times, but the citizens were equally tough. Personal liberty was fiercely defended, and governments were, by and large, too small and weak to have much say in the lives of ordinary people. These were the people whose characters built the United States, Australia, Canada and New Zealand. These were the people for whom the motto, 'Give me liberty or give me death' resonated.

The question now: 250 years later, does any of that libertarian blood still run in the veins of their descendants?

---

229 Maori warriors in New Zealand's far north had, just a few years earlier, captured and eaten a French naval captain and 26 troops. Captain James Cook, just a few years prior to that, had lost ten men to the cooking pot in another dispute. See The Great Divide by Ian Wishart, Howling At The Moon Publishing, 2012, pages 75-77

CHAPTER 14

Conquest And Rebellion

"They who can give up essential liberty to obtain a little
temporary safety deserve neither liberty nor safety."
– Benjamin Franklin

The Age of Discovery had opened the doors to the big wide world. Suddenly, European culture and colonists were spilling out of their home countries and into new lands all over the planet. Invariably they found native people already living there, and invariably there was conflict as the technologically advanced Old World took control of the new.

The Americas were close enough to Europe to allow large military attachments to be deployed in support of new settlements. Native Americans were increasingly pushed back into reservations, not just in what became the US but equally in Canada and South American colonies as well.

Much of the initial labour force in the US had come from "transportation" of petty criminals – often children – to indentured service in the new land, and their ranks were swelled by the slave trade out of Africa. Regardless of whether you were white 'servant' or black 'slave', your fate was often similar: for the women, rape and for the men flogging for even minor indiscretions.

No one was immune from transportation, not even British nobles. One such was William Parsons, the eldest son and heir to the Baronet Sir William Parsons. Like many "to the manor born", Parsons Jnr lacked the character of his father. Gifted with an education at Eton College,

Junior set the scene for his life by being caught stealing there. His crimes as a young man mounted, and eventually they caught up with him, as the *Newgate Calendar* reports:[230]

"Being convicted at the assizes at Rochester, he was sentenced to transportation for seven years; and in the following September he was put on board the *Thames*, captain Dobbins, bound for Maryland, in company with upwards of one hundred and seventy other convicts, fifty of whom died in the voyage. In November, 1749, Parsons was landed at Annapolis, in Maryland; and having remained in a state of slavery about seven weeks, a gentleman of considerable property and influence, who was not wholly unacquainted with his family, compassionating his unfortunate situation, obtained his freedom, and received him at his house in a most kind and hospitable manner.

"Parsons had not been in the gentleman's family many days before he rode off with a horse which was lent him by his benefactor, and proceeded towards Virginia; on the borders of which country he stopped a gentleman on horseback, and robbed him of five pistoles, a moidore, and ten dollars.

"A few days after, he stopped a lady and gentleman in a chaise, attended by a negro servant, and robbed them of eleven guineas and some silver: after which he directed his course to the Potomack river, where finding a ship nearly ready to sail for England, he embarked, and after a passage of twenty-five days landed at Whitehaven."

Not long back in England, Parsons Jnr was foolish enough to be caught on the highway with two pistols, cocked and loaded. Although no charges were pressed in relation to the guns, Parsons roughly matched the description of a wanted highwayman who'd been in the area recently. When authorities discovered he'd returned from the colonies seven years too early, his goose was cooked:

"Parsons was now arraigned for returning from transportation before the expiration of the term of his sentence: nothing therefore was necessary to convict him but the identifying of his person. This being done, he received sentence of death. His distressed father and wife used all their interest to obtain a pardon for him, but in vain: he was an old offender, and judged by no means a fit object for mercy."

A 1671 report by Virginia's Governor Berkeley noted that 80% of "servants" died of disease once they'd been set to work in America.[231]

---

230 The Newgate Calendar, http://www.exclassics.com/newgate/ng241.htm
231 "A People's History of the United States", Howard Zinn, Harper Collins reprint 2005, p44

Because they were white, and not as easily distinguishable from their masters as slaves, keeping track of servants required greater state surveillance. In Virginia whites were required to carry passports and other official documents at all times to prove they were "free" – something of a contradiction in terms. Anyone looking suspicious and not found with the appropriate papers was prima facie treated as a runaway servant and "delivered up".

Servants could not own property, couldn't vote, and only their masters could serve on juries. Servants were largely forbidden to marry without their master's consent and, if they did, would be punished as if they had committed adultery. In 1666, writes historian Zinn, a New England jury acquitted a couple of causing the death of a servant following a decision by the mistress of the house to chop off the servant's toes. Beatings, rapes and murders of white servants by masters were common.

Rebellious servant Isaac Friend tried to rustle up a revolt in 1661 in Virginia, urging his friends to get "gunnes".

"Who would be for Liberty and free from bondage?" he asked, posing that eternal question.

Black or white, if you were poor in the colonies you were trash. Yet even with that label, you fought for your freedom. It was this deep yearning that the leaders of the American Revolution later capitalised on, even though for the most part those revolutionary leaders were the very same "Masters" who formed America's ruling class.

In fact, ironic as it may seem, the American Revolution was like a run along the razor's edge with a strand of hair – so delicately balanced between competing factions that it was a miracle they managed to pull it off. It's a classic study in the politics of divide and rule, and a lesson in how small groups can pull off major social and political change.

What happened was this.

The process of immigration to the American colonies had made a handful of families very, very rich with huge landholdings. These people became the equivalent of lords of the manor back in Britain, owning large estates and renting small holdings out to their tenants – the middle class tradesmen and farmers. The property-owning class were the ones chiefly being pinged for taxes by the Crown, and they deeply resented the encroachment on their wealth. To get an idea for how rich and powerful they were, studies of Boston tax records in the late 1600s found that out of a total population of six thousand people, a tiny group of 50 rich men (1% of the population) owned 25% of the town's wealth. Records from

1770 show the top one percent owned a staggering 44% of the wealth just prior to the revolution.[232]

In essence then, there were three classes in colonial America: the property-owning elite, the middle classes, and the rest – slaves, servants, Native Americans and the poor.

The elite relied heavily on slaves and the servant class to provide cheap labour. By 1770, one in three adult white men in Boston owned no land and accordingly had no right to vote. The colonies were run by voting property-owners, who stacked councils and local governments with their own men. Even those white servants who'd finally been given their freedom remained disenfranchised unless, through some stroke of fortune, they could generate enough income for themselves to buy land.

At that stage, the land to the west of the Appalachian ranges was "here be dragons" territory on the map – out of bounds thanks to the large Indian populations beyond. This restriction on expansion pushed up the price of land.

The colonial elite remained fearful of the Native American tribes, and equally fearful of a potential uprising by black slaves and white servants. They were especially fearful of both those threats combining. When white servants ran away, they ran to the Indians.

"There must be in their social bond," remarked writer Hector St Jean Crevecoeur in later years, "something singularly captivating, and far superior to anything to be boasted among us; for thousands of Europeans are Indians, and we have no examples of even one of those Aborigines having from choice become Europeans."[233]

The solution to this was particularly crafty, and an epitome of the divide and rule strategy mentioned earlier. By keeping land prices high in the eastern seaboard, the colonial politicians pushed the land-seeking freed servants and poor whites out to live in the western boundaries of the territories, where they would annoy the Native Americans by building close to the Indian settlements. Not only would this aggravate the Indians and make them less tolerant of the poor whites, but it also provided a "buffer zone" of white trash between the tribes and the wealthy citizens on the coast.

A similar stunt was pulled in regard to black slaves. The administrators of South Carolina, where 25,000 whites faced 60,000 Indians and 40,000

232 Zinn, ibid, p49
233 J. Hector St. John Crèvecoeur, Letters from an American Farmer. Edited by W. P. Trent and Ludwig Lewisohn. New York: Duffield, 1904. See online at http://mith.umd.edu/eada/html/display.php?docs=crevecoeur_letters.xml

slaves, created black militia units to fight the Indians, engendering conflict between both those races. Zinn quotes one letter from the period on the policy "to make Indians and Negroes a checque upon each other lest by their Vastly Superior Numbers we should be crushed by one or the other."[234]

Even so, rebellion attempts were relatively common. Nathaniel Bacon led an uprising in 1676 in Virginia demanding the ruling class do more to protect settlers on the western frontier from Indian raiding parties. Bacon's men were indiscriminate in who they attacked – their victims included the Pamunkey tribe who'd long been allied to the settlers ever since first contact at Jamestown. In fact, Pocahontas was from that tribe. What made the rebellion more or less unique is that hundreds of black slaves and white servants joined the rebel militia. According to contemporary records, the final groups to surrender when the rebellion was finally put down were "four hundred English and Negroes in Armes", and another squad of 300 "freemen and African and English bondservants".

The absolute last thing the elite wanted was for poor whites and blacks to get a taste of freedom fighting, and especially not together as brothers in arms. As part of the divide and rule strategy, Virginia's government rewarded white servants who had joined Bacon's Rebellion with an amnesty, but continued to punish black rebels. The new laws also meant that white servants, once they became free, would be allowed to bear weapons, but free blacks would not.

The message from all this was simple: separation of people on the basis of skin colour was to become the weapon of choice in controlling society. Taking it further, free blacks were prohibited from engaging in trades or small businesses that might compete with poor and middle class whites.

Britain, meanwhile, had been embroiled in wars of its own and desperately needed tax revenue. In those days, there was no such beast as an income tax in the form that we know it. Instead, governments survived on property taxes and customs levies and sales taxes of various kinds. Those taxes were generally small (there being no sizeable civil service and no welfare, health or education agencies to soak up money), unless there was a war on and troops had to be fed. Taxes to support standing armies could be expensive.

The winds of revolution, additionally, were spreading.

---

234 Zinn, p 54

CHAPTER 15

How Governments Gained Control Of Your Money

"When the people fear the government there is tyranny,
when the government fears the people there is liberty."
– Thomas Jefferson

From Britain in the 1600s to the Americas in 1776, the concept of overthrowing the monarchy spread to France in 1789.

The US provides a salutary example of the powers at work and the issues they faced. The founding fathers wanted a system of government with a series of checks and balances, precisely to make sure that power did not concentrate in the hands of the few at the expense of many.

One of the first protections they seized on was federalism – a collection of individual states each delegating certain limited powers to a central administration. The states would retain control of their own affairs, but the central government would represent them internationally.

The US Constitution carefully spelt out that human beings had certain inalienable rights (meaning they could not be taken away by a government), but that they would delegate such limited powers as were necessary for governing to a federally-elected body. There was never any question as to who was sovereign in the American system however: "We, the People".

There was to be no state-sanctioned Church, unlike the Anglican Church in England or the Roman Catholic Church in Europe, both of which dabbled in politics. This was not a rejection of the concept of religion, however. Jefferson was a Deist[235], not a Christian, but he nonetheless saw the need for belief in something greater than the State:

"Can the liberties of a nation be thought secure when we have removed their only firm basis, a conviction in the minds of the people that these liberties are of the gift of God? That they are not to be violated but with his wrath?"[236]

Power, as Jefferson knew, loves a vacuum. When one source of authority is weakened, another rises up to fill the void.[237] A clue, perhaps, to his reasoning comes from his firm belief that a government should not try and teach you what to think – not because having moral values is wrong but because it should never be the role of good government to impose its beliefs on the people. In that sense, Jefferson would be horrified if he could see what passes for "education" these days. Here's what he said in the Virginia Statute for Religious Freedom:

"To compel a man to furnish contributions of money for the propagation of opinions which he disbelieves and abhors, is sinful and tyrannical."

In his famous Letter to the Danbury Baptists, Jefferson wrote:

"Legitimate powers of government reach actions only, and not opinions."[238]

In 1779, he wrote: "The opinions of men are not the object of civil government, nor under its jurisdiction."[239]

A government that changed society to match its own belief system was seen as totalitarian and oppressive. It could become self-perpetuating, not because it was inherently good but because it was good at indoctrination.

To prevent the possibility of a bad government taking away the freedoms

---

235 Deism is the belief in a God who created the universe but who does not intervene in human affairs, as opposed to Theism which argues that God does. Christianity, Islam and Judaism are examples of theistic faiths.

236 Jefferson, Notes on the State of Virginia, 1781 (Richmond, Va.: Randolph, 1853), 174

237 In one of Abraham Lincoln's presidential Executive Orders, a similar point is made: "Of all the dispositions and habits which lead to political prosperity, religion and morality are indispensable supports. In vain would that man claim the tribute of patriotism who should labor to subvert these great pillars of human happiness – these firmest props of the duties of men and citizens. The mere politician, equally with the pious man, ought to respect and to cherish them. A volume could not trace all their connections with private and public felicity. Let it simply be asked, Where is the security for property, for reputation, for life, if the sense of religious obligation desert the oaths which are the instruments of investigation in courts of justice? And let us with caution indulge the supposition that morality can be maintained without religion. Whatever may be conceded to the influence of refined education on minds of peculiar structure, reason and experience both forbid us to expect that national morality can prevail in exclusion of religious principle." See Abraham Lincoln: Executive Order http://www.presidency.ucsb.edu/ws/index.php?pid=69795

238 http://www.loc.gov/loc/lcib/9806/danpre.html

239 http://extext.virginia.edu/jefferson/quotations/jeff0750.htm

of citizens, the Constitution turned all citizens into soldiers, authorising the right to bear arms.

Earlier, the Constitution of Virginia contained this as a draft proposal when Thomas Jefferson wrote: "No freeman shall be debarred the use of arms [within his own lands or tenements]". A man's home, essentially, was his castle. The right to defend one's home, property and family was seen as inalienable. To defend one's Constitution and freedoms was not merely a right but a duty.

The natural state of governments, warned Jefferson in a letter to a correspondent, was to increase their powers wherever possible:[240]

"The natural progress of things is for liberty to yield, and government to gain ground."

Within decades of the Revolution, Jefferson was witnessing liberty "yield" first-hand, and complaining bitterly about young people who knew nothing of what they were giving away:

"A vast accession of strength from their younger recruits, who, having nothing in them of the feelings or principles of '76, now look to a single and splendid government of an aristocracy, founded on banking institutions, and monied incorporations under the guise and cloak of their favored branches of manufactures, commerce and navigation, riding and ruling over the plundered ploughman and beggared yeomanry."[241]

The former US president realised it wasn't just bureaucrats and politicians who posed a threat to democracy, so did merchants and bankers – people with vast financial clout:

"I sincerely believe, with you, that banking establishments are more dangerous than standing armies; and that the principle of spending money to be paid by posterity, under the name of funding, is but swindling futurity on a large scale."[242]

Private banks had begun issuing credit and charging high interest rates, and Jefferson and others were concerned at the shift of power from the sovereign's treasury to private financiers. Private banknotes were one example:[243]

"Bank-paper must be suppressed, and the circulating medium must be restored to the nation to whom it belongs."

240 Letter from Jefferson to Edward Carrington, Paris, May 27, 1788
241 Letter to William Branch Giles, 1825
242 Thomas Jefferson to John Taylor, May 28, 1816, in Ford, 10:31
243 Thomas Jefferson to John Wayles Eppes, September 11, 1813, in PTJ:RS, 6:494

According to Noam Chomsky's analysis, Jefferson quickly came to realise that vested interests were capable of effectively subverting democracy and defeating the purpose of the revolution, and that certainly seems to have been the flavour Jefferson was developing:

"I hope we shall take warning from the example and crush in its birth the aristocracy of our monied corporations which dare already to challenge our government to a trial of strength, and to bid defiance to the laws of their country."[244]

For Jefferson, in practice there were only three types of government:

"Societies exist under three forms sufficiently distinguishable. 1. Without government, as among our Indians. 2. Under governments wherein the will of every one has a just influence, as is the case in England in a slight degree, and in our states in a great one. 3. Under governments of force: as is the case in all other monarchies and in most of the other republics. To have an idea of the curse of existence under these last, they must be seen. It is a government of wolves over sheep.

"It is a problem, not clear in my mind, that the 1st. condition is not the best. But I believe it to be inconsistent with any great degree of population. The second state has a great deal of good in it. The mass of mankind under that enjoys a precious degree of liberty and happiness. It has its evils too: the principal of which is the turbulence to which it is subject. But weigh this against the oppressions of monarchy, and it becomes nothing. *Malo periculosam, libertatem quam quietam servitutem.* Even this evil is productive of good. It prevents the degeneracy of government, and nourishes a general attention to the public affairs. I hold it that a little rebellion now and then is a good thing, and as necessary in the political world as storms in the physical."[245]

Regardless of the revolution, however, times were changing; freedom was proving to be an expensive beast to feed and this would inevitably change the structure of government. Following on from the expensive American independence wars, and clashes with Spain in the early 1800s and France with the Napoleonic wars, taxes were increasingly being imposed in mother England with monotonous regularity. The Crimean campaign only added to their woes as did the New Zealand wars between British troops and rebel Maori tribes around the same time.

---

244 "Writings of Thomas Jefferson", published 1892, see http://hdl.loc.gov/loc.mss/mtj.mtjbib022651
245 Thomas Jefferson to James Madison, Paris, January 30, 1787

To maintain an empire militarily required infrastructure, an army doesn't just march on its feet but on its stomachs as well. Income tax proposals began to be dusted off as a means of paying for government infrastructure and by the 1890s most British Commonwealth colonies had tax structures in place, albeit tiny. New Zealand's first income tax, for example, was one and a half pennies for every two pound sterling of income. Hardly a major burden.

Suddenly, however, major private bankers could see the implications: they could lend money to governments, knowing the public were liable through taxation revenue to pay the bills. The concept of national debt began to take off.

Money, for centuries, had only been "worth its weight in gold", and coins were the main unit of transfer. Private banks had for a while issued their own banknotes, effectively notes promising to exchange the paper for gold or silver to a nominated value, but there was no national paper money system. Taxation changed all that.

It was a developing threat in Jefferson's day, as evidenced by his letter to Treasury Secretary Albert Gallatin in 1803 regarding a proposed central bank:

"[The] Bank of the United States... is one of the most deadly hostility, existing against the principles and form of our Constitution... An institution like this, penetrating by its branches every part of the Union, acting by command and in phalanx, may, in a critical moment, upset the government. I deem no government safe which is under the vassalage of any self-constituted authorities, or any other authority than that of the nation, or its regular functionaries. What an obstruction could not this bank of the United States, with all its branch banks, be in time of war! It might dictate to us the peace we should accept, or withdraw its aids. Ought we then to give further growth to an institution so powerful, so hostile?"

In 1820, nearly half a century after independence, Jefferson wrote again about the dangers of government borrowing:[246]

"It is incumbent on every generation to pay its own debts as it goes. A principle which, if acted on, wou [ld] save one half the wars of the world; and justifies, I think our present circumspection."

The problem for those who wanted to change the world was a simple one: who pays for it? "Give me control of a nation's money supply, and I

---

246 Jefferson letter to Antoine Louis Claude Destutt de Tracy of December 26, 1820

care not who makes the laws," boasted Meyer Amschel Rothschild, the founder of the banking dynasty, back in the late 1700s.

To compete with the banking system, politicians realised they needed an income stream of their own.

Income taxes as a regular levy were illegal under the US Constitution, although there was provision for imposing such taxes in wartime. Private bankers were charging upwards of 36% interest on loans to the US government, and US President Abraham Lincoln decided to sidestep this in 1863 by authorising the printing of $100 million in US Government banknotes with which to pay for the civil war.[247]

"That Congress has power to regulate the currency of the country can hardly admit of doubt, and that a judicious measure to prevent the deterioration of this currency, by a seasonable taxation of bank circulation or otherwise, is needed seems equally clear. Independently of this general consideration, it would be unjust to the people at large to exempt banks enjoying the special privilege of circulation from their just proportion of the public burdens."

Announcing this, Lincoln added that the time had come for America to have a national currency controlled by the Government:

"In order to raise money by way of loans most easily and cheaply, it is clearly necessary to give every possible support to the public credit. To that end a uniform currency, in which taxes, subscriptions to loans, and all other ordinary public dues, as well as all private dues, may be paid, is almost, if not quite, indispensable."

To put Lincoln's problems in perspective, his 1863 Budget statement revealed $69 million in revenue from Customs duties, $38 million from "internal revenue", and a staggering $777 million in borrowing to pay for the war.[248]

By the following year, his banking plans were well underway.

"Very soon there will be in the United States no banks of issue not authorized by Congress and no bank-note circulation not secured by the Government. That the Government and the people will derive great benefit from this change in the banking systems of the country can hardly be questioned. The national system will create a reliable and permanent

247 Abraham Lincoln: Message to Congress on Printing Money, 17 January 1863 http://www.presidency.ucsb.edu/ws/?pid=69850
248 Abraham Lincoln: "Third Annual Message," December 8, 1863. Online by Gerhard Peters and John T. Woolley, The American Presidency Project. http://www.presidency.ucsb.edu/ws/?pid=29504

influence in support of the national credit and protect the people against losses in the use of paper money," Lincoln said in his fourth Annual Message to the people, in December 1864.[249]

Together with banking reform, Lincoln was the first US president to introduce a federal income tax, initially set at 3% on incomes above $13,000 in today's money, rising to 5% on incomes above $220,000.

Once the civil war was over, the income tax lapsed, but the bureaucracy knew they had stumbled upon the secret of empire.

America did not introduce a full income tax until the 16th amendment of the Constitution was passed in 1913, but it resulted in a massive expansion of the public service. Government employees, previously few in number and limited by whatever revenue was actually in the government coffers, could now convince their masters to put spending programmes in place with taxpayers underwriting the debt. In short, the bureaucrats realised they could grow their own personal fiefdoms.

Politicians realised that if they could appeal to the poor and get their votes by promising financial relief, they could win office with a justification for boosting government spending and thus the need for "taxing the rich". In a sense, it was an easy sell, and to be fair, the rich weren't taxed nearly as highly as they are today.

The significance, however, was global.

For the first time in human history, governments everywhere had stumbled upon a means of forcing citizens to pay them on a regular basis. Income tax is a modern invention, barely a century old. In paying for essential and non-essential government services however, income tax has enabled an explosion in the bureaucratic classes, and an equivalent turbocharging of their powers.

There was a shift in power and resource as people stopped working for themselves and began working for the state. They owed their living to the State and, in turn, they devised or worked on schemes that made a growing number of citizens dependent on the State at some level.

Across the planet, taxation revenues have paid for bureaucrats and politicians to attend international talkfests, conferences and meetings to thrash out common goals and objectives and to be briefed on new ideas in a range of areas. Taxation revenues have paid for national school systems

---

249 Abraham Lincoln: Fourth Annual Message http://www.presidency.ucsb.edu/ws/index.php?pid=29505

to train generations of youth in whatever doctrines their State educators have endorsed. Taxation revenues have allowed governments to go into hock to private banks and borrow vast sums of money on the promise of levies on the incomes of taxpayers not even born yet

In the space of just 300 years since the first British settlers arrived at Jamestown, America and the world had managed to find the financial secret that would make control of the world possible, for those who knew how to manipulate the system.

Who were those people and what was their motivation? Let's find out.

CHAPTER 16

The Enlightenment: Rise Of The Illuminati

"He who controls the past controls the future. He who controls the present controls the past."
– George Orwell, 1984

Behind all the revolutionary activity in Britain, the USA and France lay something much deeper. In popular historical jargon, it's known as "The Enlightenment", a period of intellectual growth in both discovery and knowledge that set humankind on a new pathway. The word "enlightenment" was not coined by accident, but by a group of mystics and freemasons for whom the word had a hidden spiritual meaning.

Prime movers behind the Enlightenment (also referred to as the Age of Reason) included mystic philosopher Baruch Spinoza, who argued heavily against the Catholic Church and claimed in his writings that everything – every human, every plant, every animal – was part of 'god'. The technical name for this belief is 'panentheism', literally, "everything in god". Spinoza didn't see his god as a personality, but just a thing, a force, that cared no more for humanity than for anything else on the planet.

The appeal to western intellectuals was that it gave them a belief system free of the trappings of Christianity with its emphasis on a personal God. The panentheistic force, on the other hand, judged no one and was simply there to be tapped into. Back in the 1700s, they called Spinoza's beliefs atheism, and he called it rationalism. The poets Shelley and Coleridge called Spinoza's ideas "a religion of nature"[250]. In the 1970s, we'd be calling this philosophy Shirley Maclaineism, aka the New Age. In the twenty-teens we're more likely to call it "the Green Religion".

The idea of a divine force governing nature began to underpin scientific thinking. Others to leap on board the Enlightenment bandwagon included British philosopher John Locke, scientist Isaac Newton and French philosopher Francois-Marie Voltaire.

Locke believed in the Christian God and much of the Bible, but believed divine law and natural law were intertwined. He was big on religious tolerance, made a strong case for the equality of women based on the Genesis story of Adam and Eve and both being created in the image of God, and his writings influenced Thomas Jefferson and other patriarchs of the American Revolution. It was from Locke that the Declaration of Independence drew its "All men are created equal, ... they are endowed by their Creator with certain unalienable rights, ... life, liberty, and the pursuit of happiness."

Voltaire, on the other hand, was a strident anti-Christian:

"The establishment of Christianity [was] a grievous aberration of the human mind, a halt in the progress of humanity".[251]

Like Spinoza, Voltaire's theology leaned towards nature religion, and he saw Christ as a "master" but not a saviour. Like Locke, Voltaire's writings on religious tolerance influenced Jefferson.

As the opposition to classical Christianity grew, so too did the rise of freemasonry. This secretive sect would have you believe they can trace their origins back to the ancient Egyptians. In reality freemasonry might draw on ancient religious influences, including a claimed link to King Solomon's temple, but the Lodges as we know them today were born of the Jacobite era in Scotland and England during the 1600s and 1700s.

The essence of freemasonry was simple, however: they were a society of men who pledged loyalty to the brotherhood, and whom were initiated into a series of veiled occult rites. Freemasonry claims to be secular, but that's the wrong word to use. Non-denominational is a better description. Allegiance is given to "the Great Architect of the Universe", whose 'wisdom' is slowly revealed to Masonic initiates as they grind their way up the scale of "degrees".

250 "Spinoza: A Life" by Steven Nadler, Cambridge University Press, 1999
251 From Voltaire's 1762 work on rebel priest Jean Meslier

The Catholic Church outlawed freemasonry and excommunicated any of its flock known to be masons. Freemasonry's response was to hide its membership behind secret handshakes and symbols. This played on a kind of snobbery that underpins freemasonry and its modern equivalents – a belief that the secrets of the world and the universe are hidden from ordinary men and women and only revealed to those specially chosen, an 'elite'. This spiritual pecking order gave a natural fuel to the degrees programme and lent a kind of "if I told you, I'd have to kill you" pathos to the whole enterprise. Today, for example, many common or garden-variety freemasons don't have the foggiest idea what their organisation truly stands for, because they simply are not high enough up the ladder yet.

In fact, freemasonry is merely an old cult known as "Gnosticism", from the Greek word for knowledge, *gnosis*. The opposite word "agnostic" simply means, "I don't know".

The Gnostics sprung up from a source unknown but became most prominent around a century after the death of Jesus Christ. Their basic belief was in a confrontation between spirit and matter, which they saw as good and evil. The material world was bad, the spiritual world was good, and humanity was in a state of darkness until individuals became "enlightened" by supposed spiritual secrets.

Followers of Gnosticism were particularly enraged by the Christian doctrine that Jesus Christ was God incarnate who took on a material human form to absorb humanity's spiritual sin, die on the cross and be resurrected. According to the Gnostics, it was the fate of most humans to perish in their ignorance and only an elite few would survive to the next life, not because of Jesus Christ but because of their secret knowledge.

When the Gnostics realised they were rapidly losing ground to Christianity, they stopped openly fighting Christianity and instead tried to plagiarize its main doctrines and rewrite them as Gnostic gospels. The so-called Gospel of Thomas is just one example of the Gnostics borrowing Christian themes and manipulating them to tell a different story.

Gnosticism is at the heart of freemasonry and the New Age movements, and it was at the heart of the Enlightenment movement that overshadowed the American and French revolutions.

One of the pioneer thinkers on economics and the free market, for example, was Adam Smith. Born in 1723 in Scotland, the young Smith was kidnapped by gypsies at age four and had to be rescued. Later in childhood he began having long conversations with imaginary friends and had a reputation for absent mindedness that lasted his entire life.

Smith is studied in universities today for his books like "The Wealth of Nations" and "The Theory of Moral Sentiments". You might remember Smith for his reference to "the invisible Hand" guiding market economics. Many scholars have labelled him a Christian capitalist, but Smith repeatedly made reference in his books and letters to "the Great Architect of the Universe". He may have been a cultural Christian in the sense he was born into a nominally Christian society, but spiritually Smith was a freemason.

He linked up with French philosopher Voltaire and Scottish atheist David Hume. Another of Smith's contemporaries was Adam Weishaupt. This Bavarian-born philosopher became an enlightenment guru as well.

Weishaupt's goal was to establish what he called a brotherhood of humanity founded on enlightenment principles, with equality for all worldwide and a new world order. A noble goal, but often the road to hell is paved with good intentions, and it was in this case as well.

After being disillusioned by the relative primitivism of freemasonry in 1774, Weishaupt vowed to find another way and, on 1 May 1776 – just weeks ahead of the US Declaration of Independence – Weishaupt founded a secret society of his own called the Order of Perfectibilists. You might know it better by the name its members gave themselves: the Illuminati. The 'enlightened ones'. The date of 1 May was carefully chosen, it was the occult festival Beltaine.

"His order was to be based entirely on human nature and observation," reports the *Catholic Encyclopedia* entry on the Illuminati, "hence its degrees, ceremonies, and statutes were to be developed only gradually; then, in the light of experience and wider knowledge, and with the co-operation of all the members, they were to be steadily improved. For his prototype he relied mainly on Freemasonry, in accordance with which he modelled the degrees and ceremonial of his order.

Weishaupt had no illusions that his plan for world domination would ultimately be democratic, but he knew he had to sell the dream that way to the unwashed stupid masses. Documents seized by Bavarian prosecutors when they raided Illuminati members in the 1780s spelt out Weishaupt's plans for the totalitarian "despotism of superiors" over the "blind, unconditional obedience of subordinates".[252]

---

252 PHILO (the Illuminati code name of member Freiherr von Knigge), *Endliche Erklärung* (1788)

In 1777 Weishaupt re-entered a Masonic lodge with the intention of converting the members to Illuminism. It was slow progress at first, then in 1780 one of freemasonry's leaders, Freiherr von Knigge, who used the code-name "Philo", endorsed the Illuminati, and the race was on. Within two years, more than 500 men had joined.

The Illuminati were so secretive, however, that most members only knew their immediate superiors in the organisation, and perhaps a couple of others. Very few knew that Adam Weishaupt was at the top of the tree, for example. The Illuminati in a sense set the blue print for al Qa'ida's cellular terrorist structure in the 2000's. That wasn't the only thing Osama bin Laden and Adam Weishaupt shared in common. Both had an ambition to create a new world to their liking.

As Weishaupt wrote, salvation of mankind would come from what he called secret wisdom schools to train initiates, and such schools "were always the archives of nature and of the rights of man; through their agency, man will recover from his fall; princes and nations, without violence to force them, will vanish from the earth; the human race will become one family, and the world the habitation of rational beings. Moral science alone will effect these reforms 'imperceptibly'; every father will become, like Abraham and the patriarchs, the priest and absolute lord of his household, and reason will be man's only code of law".[253]

Except it isn't reason. The Illuminati and Freemasonry might have invoked many things, but rationality genuinely was not one of them. Both organisations aimed to spread an old pagan religion deep into the heart of respectable society, and cloak it in so much quasi-Christian garb that people would gladly swallow it, as if it were a sugar pill. Their disciples were sent into the realms of science, education, business and religion, and told to report back to their superiors with details of the contacts they were making.

"The purpose of this and other regulations," reports the *Catholic Encyclopedia*, "was to enable the order to attain its object by securing for it a controlling influence in all directions, and especially by pressing culture and enlightenment into its service. All illuministic and official organs, the press, schools, seminaries, cathedral chapters (hence, too, all appointments to sees, pulpits, and chairs) were to be brought as far as possible under the influence of the organization, and princes themselves were to

---

253 "Nachtrag", by Adam Weishaupt, 1787, p80

be surrounded by a legion of enlightened men, in order not only to dis-arm their opposition, but also to compel their energetic co-operation. A complete transformation would thus be effected; public opinion would be controlled; "priests and princes" would find their hands tied; the marplots who ventured to interfere would repent their temerity; and the order would become an object of dread to all its enemies."

Weishaupt, ironically, got booted out of the Illuminati, and ended up becoming a devout Catholic again, but not before the organisation had well and truly spread its tentacles into Europe and America, through freemasonry, the churches and high society.

Thomas Jefferson, for example, was familiar with the Illuminati leader thanks to a book by one of Weishaupt's critics, and in a letter to the Reverend James Madison approaches the topic as if both are intimately aware of freemasonry (the 'craft'):[254]

"Wishaupt [sic] seems to be an enthusiastic Philanthropist. He is among those (as you know the excellent [Richard] Price and Priestley also are) who believe in the indefinite perfectibility of man. He thinks he may in time be rendered so perfect that he will be able to govern himself in every circumstance so as to injure none, to do all the good he can, to leave government no occasion to exercise their powers over him, & of course to render political government useless.

"As Wishaupt lived under the tyranny of a despot & priests, he knew that caution was necessary even in spreading information, & the principles of pure morality.... He proposed to initiate new members into his body by gradations proportioned to his fears of the thunderbolts of tyranny. This has given an air of mystery to his views, was the foundation of his banishment, the subversion of the masonic order, & is the colour for the ravings against him of Robinson, Barruel & Morse, whose real fears are that the craft would be endangered by the spreading of information, reason, & natural morality among men."

Jefferson, in typical Masonic language[255], referred in the same letter to "our Grand Master Jesus of Nazareth". His approval of Weishaupt's ideas is seen in his equation of them with those of the "excellent" Richard Price, another illumined thinker and freemason.

Just how far Masonry influenced the US revolution has been a matter

254 Letter from Thomas Jefferson to Rev. James Madison, 31 January 1800, http://memory.loc.gov/
cgi-bin/query/r?ammem/mtj:@field(DOCID+@lit(tj090050))
255 http://www.mastermason.com/dresden/thomas_jefferson.htm

of considerable debate. George Washington and Benjamin Franklin were both freemasons. At the heart of many arguments has been the design of the 1782 Great Seal of the United States, featuring on its reverse side the "all seeing eye" atop a pyramid with the phrase, "Novus ordo seclorum". This Latin inscription has been loosely translated as "New World Order", but in the finer details of Latin actually means "New Order of the Ages" if you wish to be pedantic. This seal appears on the US banknotes.

Apologists for freemasonry have argued that the all-seeing eye first appeared in Masonic usage in 1797 in a book of symbols and that it therefore followed the US Great Seal and did not inspire it, but there are several logical and factual flaws with that argument. Firstly, its appearance in a collection of Masonic symbols in 1797 implies it was being used by Masons prior to that, in order to have made the cut for the book as a commonly-accepted symbol.

Evidence of this is clearly seen in the Masonic apron gifted to George Washington in 1784 and now on display in the museum of the Masonic temple of Philadelphia. The apron prominently contains the image of the all-seeing eye. Given its production in 1784, this again implies that the symbol had been used by masons for some time prior to that.

Secondly, the Declaration of Human Rights for the French Revolution of 1789, also on museum display, prominently has the image of the all-seeing eye ensconced right at the top of the Declaration. Was there a link between the French Revolution and freemasonry? We shall find out shortly.

You can see, then, that freemasonry and Adam Weishaupt's Illuminati principles stretched as high as the Presidency of the United States, although Jefferson did draw a distinction between the necessary secrecy of the Illuminati in Europe, and the relatively open way the same ideas were being discussed in America:

"If Wishaupt [sic] had written here, where no secrecy is necessary in our endeavors to render men wise & virtuous, he would not have thought of any secret machinery for that purpose."[256]

The Enlightenment and Weishaupt's 'new world order' was also a driving force in the 1789 French revolution. The events of the Revolution are vividly captured in the epic 1859 three volume work by historian Thomas Wright, "The History of France". Wright explains how the Marquis of Mirabeau became a key player in stoking the fires of the French Revolu-

---

256 Jefferson letter to Rev. James Madison, supra

The design of the Great Seal of the United States went through a number of drafts before culminating in the image at the bottom for its reverse side. The one common feature is the all-seeing eye.

The all seeing eye features in the Illuminati/Masonic inspired Declaration of Rights for the French Revolution in 1789, and of course on another of the preliminary designs for the US Great Seal

tion. The book also notes how Mirabeau had "sold to the booksellers a secret correspondence relating to the court of Berlin, the publication of which had caused so great a scandal that the book was seized, and condemned to be burnt."[257]

That scandalous book was Mirabeau's "Secret History of the Court of Berlin", published with the assistance of a Bavarian Illuminati member named Friedrich Nicolai and which caused red faces in both Berlin and Paris. Mirabeau was also introduced to the Duke of Brunswick, one of the highest-placed royal Illuminati, and a man who had pledged the resources of the Templar freemasonry lodges in 1782 to ushering in a French republic.

Within months of the Berlin book Mirabeau was at work on another, "On the Prussian Monarchy", which among other things set out the Illuminati's plans for the overthrow of the French monarchy. With no radio or TV stations, publicity was gained by the mass printing of books and pamphlets, and the resulting distribution of them on the streets. He who controlled the war of ideas controlled the crowd.

Mirabeau was assisted in this by the Duke of Orleans, Philippe Egalité, and his colleague Abbé Joseph Sièyes. Both men were, like Mirabeau, extremely highly-placed freemasons. Sièyes, in particular, was regarded as one of the most prolific pamphleteers of the revolution, and went on to become an advisor to Napoleon Bonaparte in his efforts to spread the French revolution across Europe.[258]

Add to the mix Swiss bankers and freemasons Panchaud and Clavière who were bankrolling Mirabeau's efforts, and the makings of a perfect revolutionary storm were there.[259] The group hatched a cunning plan to make themselves a vast fortune (with which they could pay for more propaganda) and at the same time near-bankrupt the treasury of the King of France. They set in motion a stockmarket crash, and played the King's treasurer for a sucker.

"The affairs of the Revolution everyday are going better in France," wrote Illuminati member Jakob Mauvillon – a protégé of Weishaupt's

257 "The History of France, Vol II" by Thomas Wright, 1859, p423
258 The French commander La Fayette, who played a big role assisting George Washington in the War of Independence, was a freemason and presented fellow mason Washington with a Masonic apron. La Fayette was likewise an instrumental figure in the later French Revolution. See http://www.pagrandlodge.org/mlam/apron/
259 This is not to suggest that the Revolution was entirely and solely the work of freemasons or the Illuminati. Far from it. The French monarchy had done enough to discredit itself already, the opportunists simply seized the day and made the most of it.

and a mentor to Mirabeau. "I hope above all that within a few years that the flame will take hold also and be embraced generally, then our Order will be able to do grand things."[260]

The essential themes of the revolution – liberté, egalité, fraternité – were the themes of Weishaupt's order: freedom, equality, brotherhood. You couldn't change the world unless you brought the crowd with you, and you couldn't bring the crowd with you if you didn't give them something to aim for.

Of course, this wasn't really any different from the doctrines of Christianity either, but that's where the similarities ended. Freemasonry and the Illuminati represented a very different spiritual entity – the mysterious "Grand Architect of the Universe" whose true purpose had only been revealed to the absolute highest order of initiates. You will meet the "Grand Architect" in due course in this book.

Central to all these struggles was an overwhelming belief in the need for some kind of new world order. The masons, the Illuminati and the revolutionaries all believed they were on the verge of ushering in a new era in human history, an age no longer dominated by belief in human salvation by God, but by human salvation by the acts of men. Humans, they believed, were the authors of their own destinies, and would become god-like the closer they got to moral perfection.

The argument may have been couched in political terms, but the underlying, driving force was a spiritual one, a belief in a higher purpose. Politics was merely the mode of transport to take people towards that goal.

Like many revolutions in their time, the Illuminati were merely a brief temporal manifestation of a much more eternal idea. The organisation known as the Illuminati was absorbed within the wider freemasonry movement by the 1820s, where its seeds found fertile ground. As previously indicated, founder Adam Weishaupt had long since abandoned it and returned to orthodox Catholicism. The ideas he unleashed, however – a re-education of the public towards a new morality and a new world order – were to become the foundation stones of the creep towards a totalitarian global regime.

If anyone today tells you the Illuminati exist as an actual order, they're probably wrong. On the other hand, the concept of an "enlightened" group of elite is certainly still with us, and the Illuminati agenda laid out by Adam Weishaupt is well and truly in play, as we are about to see.

---

260 "Les Illuminés de Bavière et la Franc-Maçonnerieallemande" by René Le Forestier, Paris, 1915, p663

CHAPTER 17

The Rise Of The United Nations

*"Always think far into the future, for the seeds of
tomorrow are being planted today"*
*– Dr Robert Muller,
Asst. Secy. General, United Nations*

World War One, with its origins in the collapse of the Austro-Hungarian
and Ottoman Empires, and the ambitions of a united Germany, had a
massive impact on the civilised world. Millions had died in the war,
millions more from the global Spanish flu epidemic that followed as the
troops came home in overcrowded and disease-ridden transport ships.

The histories of the twentieth century's two great confrontations have
been extensively covered by others, and need no detailed repetition here.

Weary of battle, world leaders established the League of Nations – with
the notable absence of the USA – in the hope of settling future disputes
without bloodshed. The United States came on board in a different way
a decade later when it signed the Kellogg-Briand Treaty for the Renunciation of War in 1929:[261]

"I congratulate this assembly," said US President Herbert Hoover, "the
states it represents, and indeed, the entire world upon the coming into
force of this additional instrument of humane endeavor to do away with
war as an instrument of national policy and to obtain by pacific means

---

261 Herbert Hoover: Remarks Upon Proclaiming the Treaty for the Renunciation of War (Kellogg-
Briand Pact). http://www.presidency.ucsb.edu/ws/index.php?pid=21873

alone the settlement of international disputes… an act so auspicious for the future happiness of mankind has now been consummated. I dare predict that the influence of the Treaty for the Renunciation of War will be felt in a large proportion of all future international acts."

Key signatories to the Treaty, apart from the US, included Britain, Australia, New Zealand, Japan, Italy and Germany. Within ten years, despite a "binding" declaration of peace, a war the size of which the world had never seen was breaking out as the planet went through a military version of groundhog day.

From 1939 to 1945, virtually the entire globe was involved in conflict at some level.

In the ashes of World War 2, world leaders again looked for ways to end such horrors. Clearly, ill-fated treaties like the Kellogg-Briand Renunciation of War were not going to cut the mustard any longer.

The mantle of world policeman had passed from Great Britain to the United States, whose late entry into WW2 had proved decisive. The League Of Nations, established after WW1, had obviously proven ineffective at preventing the rise of Adolf Hitler and the re-armament of the German state; something stronger was needed.

The global conflict of 1939-1945 had left virtually no corner of the world untouched. Of necessity, almost every sovereign power on the planet had been forced to choose sides and work within international treaties and agreements to fight in cooperation with other like-minded countries.

The doors of international cooperation opened by the war on the Nazis could not be closed again. The Allies had demonstrated that international alliances could be truly effective in policing the globe. One of the first symbols of the new hope to arise was the United Nations.

Leaders and diplomats met in San Francisco late 1945 to thrash out the fine print for the new organisation. As President Obama recently described it:

"After the Yalta Conference, shortly before the United Nations was founded, President Roosevelt spoke of what it would take to achieve a lasting and durable peace. 'The structure of world peace,' he said, 'cannot be the work of one man or one party or one nation. It cannot be an American peace or a British, a Russian, a French, or a Chinese peace. It cannot be a peace of large nations or of small nations. It must be a peace which rests on the cooperative effort of the whole world'.

"These words," said Obama, "are more true today than ever, whether it's

Robert Muller

preventing the spread of deadly weapons, promoting democratic governance, or fighting daily battles against poverty and hunger and deprivation. International peace, development, and security will not be achieved by one nation or one group of nations. It must be the work of all of us."[262]

Roosevelt, the first person to use the phrase "United Nations" to describe the Allies during WW2, died on the eve of the establishment of the UN with a sudden brain haemorrhage – it was early April 1945. By the time the diplomats and leaders signed off on the UN Charter in October that year, the war had ended with the surrender of Japan.

In the decades since, there has been huge support internationally for the United Nations concept, and what it may lead to. French leader Charles de Gaulle, for example, ominously remarked, "Nations must unite in a world government or perish."[263]

"It seems to many of us that if we are to avoid the eventual catastrophic world conflict, we must strengthen the United Nations as a first step toward world government," opined US newsman Walter Kronkite, "and [empower]police to enforce its international laws and keep the peace... To do that, of course, we Americans will have to yield ... It would take a lot of courage, a lot of faith in the new order."[264]

You can see, there, why the news media are so sycophantic to the UN agenda, it's in their blood.

The UN might have been born as a political organisation, but few will know that one of its driving forces was a New Age lawyer, Robert Muller. Born in Belgium, March 11, 1923, Muller had known the horrors of the Nazis as a resistance fighter. His experiences led him to write an essay on how to govern the world, and that essay secured him an internship at the

262 http://www.whitehouse.gov/the-press-office/2013/09/24/remarks-president-obama-luncheon-hosted-un-secretary-general-ban-ki-moon
263 "Peace Through World Government", George W. Blount, Moore Publishing Company, 1974, p.30
264 "Neo Gnostics at the End of the Age" by Mary Jo Anderson, Catholic Culture, http://www.catholicculture.org/culture/library/view.cfm?recnum=4635

UN in 1948. The topic had been chosen by the UN itself, so its agenda has been reasonably plain from its inception.

Muller quickly rose through the ranks as one of the UN's top bureaucrats. In a 40 year career he was at the centre of forming most of the UN's big-ticket policies, as UN Secretary-General Ban Ki-moon noted in a 2011 eulogy:[265]

"His creativity and influence were instrumental in the conception of scores of multilateral bodies, including the UN Development Programme, the World Food Programme, the UN Population Fund, and the World Youth Assembly – some of the key endeavours that define the United Nations and our global work. He knew that these could not be impersonal institutional structures but, rather, had to be alive with vision, compassion and a powerful sense of human unity.

"Dr. Muller brought precisely that spirit to whatever task he undertook. For him, the entire human family was his family. He also helped orchestrate the first-ever UN Conference on the Human Environment in Stockholm in 1972, and was deeply involved in many other global gatherings. His career spanned the decades, the issues, and even almost all the professional ranks of the Secretariat – starting from the P1 level and rising steadily to Assistant-Secretary-General, directly serving three of my predecessors. He also had a tremendous imprint on global education, including through the University of Peace, and richly earned the UNESCO Prize for Peace Education in 1989."

It was Muller who gave us the climate change bandwagon and whose philosophies infused the global "peace" movement. He was the man who pushed to enlist the world's youth in UN affairs (for reasons which will shortly become clear). But who, exactly, was this Rasputin? He didn't see himself as a diplomat, he saw himself as an evangelist for the New Age, and was directing the United Nations to a position of global supremacy so it could usher in a new spiritual power to govern the planet. I kid you not. Here is Muller endorsing the words of his spiritual advisor Sri Chinmoy:[266]

"No *human* force will ever be able to destroy the United Nations, for the United Nations is not a mere building or a mere idea; it is not a man-made creation."

Really? According to Muller and Chinmoy, the United Nations is

---

265 "New York, 11 March 2011 – Secretary-General's remarks at Memorial Service for Robert Muller", http://www.un.org/sg/statements/?nid=5132
266 "My Testament To The UN" by Robert Muller, World Happiness and Cooperation, USA, 1994, p172

supernaturally-driven, so if you think of it as an independent, secular institution you are in for a shock:

"The United Nations is the vision-light of the Absolute Supreme, which is slowly, steadily and unerringly illuminating the ignorance, the night of our human life.

"The divine success and supreme progress of the United Nations is bound to become a reality. At his choice hour, the Absolute Supreme will ring His own victory-bell here on Earth through the loving and serving heart of the United Nations."

Who knew the UN was "divine" in nature, ordained by "the Absolute Supreme"?

In case you are wondering who the coy phrase "Absolute Supreme" might refer to, don't get distracted, we will get to that shortly.

Muller and Sri Chinmoy are not the only senior UN figures to have invoked the concept of "the Divine" in regard to the mission of the United Nations. You'll recall at the start of this book we talked about a "convergence of opportunities", and different groups having different motives whilst sharing a similar goal.

What you are about to read will stun you. You might have heard about parts of it on some conspiracy site on the web somewhere, or read about it in some rabid, faded tome from the 1970s written by some guy claiming the end of the world was coming. Let's face it, we've all heard conspiracy theories about the United Nations.

So here's the thing. As an investigative journalist, it's my job to chase down rumours and gossip and look instead for hard facts. What follows has come directly from United Nations or other primary source documents (Muller's archives for instance). What you are about to read is the story of how an obscure pagan religious group seized an opportunity to influence the influencers, to evangelise to the power brokers, in their own attempt to control the world.

In the Prologue to this book we examined convergence of opportunity, the idea that times arise when different groups see an opportunity to cooperate towards a common purpose. Some in this story are motivated by something spiritual, while for others it is money and power.

What you are about to read is the story of the totalitarians motivated by their own brand of spiritual fundamentalism, and you will be able to judge for yourself just how successful they've been.

Let's begin.

CHAPTER 18

Iron Fist, Velvet Glove

"They pretended, perhaps they even believed, that
they had seized power unwillingly and for a limited
time, and that just around the corner there lay a
paradise where human beings would be free and
equal. We are not like that. We know that no one
ever seizes power with the intention of relinquishing
it. Power is not a means; it is an end. One does
not establish a dictatorship in order to safeguard
a revolution; one makes the revolution in order to
establish the dictatorship. The object of persecution is
persecution. The object of torture is torture. The object
of power is power. Now you begin to understand me."
–George Orwell, 1984

The UN's second Secretary-General, Dag Hammarskjöld, noted shortly
before his death in a plane crash, "I see no hope for permanent world
peace. We have tried and failed miserably. *Unless the world has a spiritual
rebirth*, civilisation is doomed."[267] (emphasis added)

It's a very odd comment for the head of the United Nations to say,
wouldn't you think? The existence of one New Age think-tanker inside

---

267 Quoted by Muller in his education essay "Of Good Teachers II", reprinted in "From What War
Taught Me About Peace" by Muller, Doubleday NY 1985, http://robertmuller.org/pdf/RobertMuller-
GlobalEducation.pdf

the UN, in the form of Robert Muller, you could explain away, but two? This guy was the head of the UN.

At the heart of the United Nations building in New York is a room with a stone altar. Not just any old lump of stone, but a 6,500 kilogram block of crystalline iron ore (magnetite) taken from a Swedish mine and illuminated by a single beam of light from a hidden source. It was Hammarskjöld who put it there and these are his words in an interview:

"The significance of a room is not the walls but is in what is framed by the walls; that is to say, we had to create a room of stillness with perhaps one or two very simple symbols, light and light striking on stone. It is for that reason that in the center of the Room there is this block of iron ore, glimmering like ice in a shaft of light from above. That is the only symbol in the Room – a meeting of light of sky and the earth.

"However, in a certain sense the symbolism goes one step further. I do not know whether there is anything quite like the arrangement of that Room with a big block of stone in the center. The original idea was one which I think you will all recognize; you will find it in many great religions; it is the empty altar, empty not because there is no God, but empty because God is worshipped in so many forms. The stone in the center is the altar to the God of all."[268]

The Secretary-General's work on the design of the room was extremely precise:

"Among the many crises he had to contend with," reported one New Age writer,[269] "he took on as a special personal project, the enlargement and reconstruction of this room. For example, he is known to have visited the room at 2 a.m. in the morning supervising the painters to make sure that the light in the room would be to his specifications. In his private correspondence that can be found in the United Nations Research Library, one letter refers to the floor plan, which as he wrote 'has been described in detail and marked with red lines.' And, for the sake of precision, he writes that he gave the dimensions in centimetres. He even included a memo for some weaving instructions for the rug that was to be placed in a very specific part of the room.

---

268 "A Brief Overview of the Spiritual History of the United Nations", UN NGO Committee on Spirituality, Values and Global Concerns, http://www.csvgc-ny.org/content/view/17/36/
269 "The United Nations Meditation Room", *Diamond Light – Newsletter of the Aquarian Age Community*, 2007 No. 3 & 4, http://www.aquaac.org/dl/07nl34art3.html

"As with any truly occult or spiritual symbol, this room represents more than meets the eye. It is replete with significant symbols, and the numerology, astrology and sacred geometry that are all a part of this very small room radiate their own silent but powerful spiritual vibrations, invoking the Soul of those who seek to work through this organization for the benefit of humanity and the planet."

At the heart of the United Nations building, then, lies a magnetic stone altar illuminated by a single shaft of light, dedicated to the "God of all". Who is this "God of all"? Is it the same entity Muller talked of in his UN writings as "the Absolute Supreme"? Why would the United Nations, of all organisations, put some kind of iron stone religious *altar* in the heart of its headquarters? Who is in charge of this altar?

Answers shortly.

So, we've established this altar was constructed on the order of the second UN Secretary-General, Dag Hammarskjöld, Robert Muller's boss. The Swede was killed in a plane crash, however, and Muller's go-to guru became Dag Hammarskjöld's successor, Burmese teacher and spiritualist U Thant who became the third person to hold top office in the UN, beginning 1961. Thant, in a "speech to Planetary Citizens" talked of the four qualities of humanity: physical, mental, moral and spiritual. "I would rate spiritual qualities the highest," said Thant.[270]

Of his mentor, Muller writes: "U Thant believed that peace on earth could be achieved only through proper education of the younger generations and that spirituality deserved the highest place in such education."[271]

Thant's vision was a future world governing federation, where the United Nations was the supreme authority and individual nation states had limited powers of self-government, similar to those of the individual American, Canadian or Australian states.

"World federalists hold before us the vision of a unified mankind living in peace under a just world order... The heart of their program – a world under law – is realistic and attainable," said Thant.[272]

This new system might look like any other political system, but UN bureaucrats have been indoctrinated with the UN belief system. Their

---

270 reprinted in "From What War Taught Me About Peace" by Muller, Doubleday NY 1985, http://robertmuller.org/pdf/RobertMuller-GlobalEducation.pdf
271 Ibid
272 "Let's Abolish War" by Tom Hudgens, citing U Thant, (Denver, Co: BILR Corp, 1986 Edition), p. 41; We the People of the World (World Federalist Association), p. 6

ABOVE: The mysterious iron stone altar to the "God of all" in a specially designed room to precise dimensions, immediately below the General Assembly room. The iron block stands 1.5m (5ft) high. BELOW: UN Secretary-General Dag Hammarskjöld organised for the largest block of iron ore ever mined to be shipped to the UN building as the centrepiece of the Meditation Room. He was killed in a plane crash.

To bring "to light the hidden things of darkness."

EDITED BY

. P. Blavatsky and Mabel Collins

The Light-bearer is the Morning Star or Lucifer; and "Lucifer is no profane or Satanic title. It is the Latin *Luciferus*, the Light-bringer, the Morning Star, equivalent to the Greek φωσφόρος the name of the pure, pale herald of daylight."—Yonge.

TOP LEFT: The Theosophical Society Journal, "Lucifer". TOP RIGHT: U Thant greets John F Kennedy at the UN. BELOW: The United Nations building in Manhattan.

beliefs on how things should be done, and even what we should all believe, will dominate any world federation.

Robert Muller saw his job within the UN as a forerunner, paving the way for the regime to come. As he indicated in the quote at the start of this chapter, this was a long game, not a short one, and to win such a long game one began with children.

In 1989, Muller talked of why he was getting the United Nations policies taught in schools around the world as part of the "World Core Curriculum" project. It was, he said, necessary to re-educate the children if he wanted to change the world:[273]

"UN Secretary General U Thant, a former school headmaster from Burma, and my spiritual master, often said to me when I was his assistant: 'Robert, there will be no peace on Earth, if there is not a new education'. He was right."

Muller and Thant were not the first to realise that capturing the future meant capturing the youth.

"When an opponent declares, 'I will not come over to your side', I calmly say, 'Your child belongs to us already. A people lives forever. What are you? You will pass on. Your descendants, however, now stand in the new camp. In a short time they will know nothing else but this new community."

The man who uttered those words was Adolf Hitler.[274]

"This new Reich will give its youth to no one, but will itself take youth and give to youth its education and its own upbringing."

The post war official investigations into the Nazi regime quickly identified just how big a role the state education system had played in preparing the German people to support Nazi ideology:

"The nation was being psychologically prepared," one tribunal heard. "One of the most important steps was the re-shaping of the educational system so as to educate the German youth to be amenable to their will."[275]

The Communist movement, likewise, elevated the importance of State education in brainwashing to create compliant and willing believers in State policies. Leon Trotsky wrote of "concentrating the education of the new generations in the hands of the state," hoping to thereby "separate

273 Acceptance speech by Robert Muller, Laureate of the UNESCO Prize 1989 for Peace Education, http://www.goodmorningworld.org/dreams/001/index.htm
274 The Nizkor Project, 23 November 1945 testimony, http://www.nizkor.org/hweb/imt/tgmwc/tgmwc-01/tgmwc-01-04-01.html
275 ibid

the children from the family...a shaking of parental authority to its very foundations."[276]

In a 1949 follow-up, Marxist educationalist Antonio Gramsci wrote of the need to ensure the education system not only taught the 3 'R's – reading, writing and arithmetic – but also taught children about their "rights and duties" to the State. Education was no longer about bettering the individual, it was about turning the individual into a follower of State doctrine.

Which brings us back to the United Nations and Dr Robert Muller's global education framework. He told of his dream "that all schools of this Earth will teach about the United Nations, which is the young people's greatest hope and will be their instrument of global action when they are grown up."[277]

For those outside of a teacher's reach, Muller knew he could rely on UN supporters in the news media and said that those in "the media who have a major role as educators will follow the example of Ted Turner[278] and will inform, teach, illustrate and make audiences participate in the building of a better world. In particular, it is imperative that they inform the public of the world information, achievements and constructive work of the United Nations system."

Just look at how the news media fall over themselves to breathlessly report the UN's every utterance on climate change and the need for a global treaty, and you can see the indoctrination in action, "full details at six".

In his essay on the World Core Curriculum, UN Assistant Secretary-General Robert Muller made his position very clear:

"If...we assume that some cosmic force or law or god or creator in the universe has put in the human species certain objectives, functions, expectations, and destinations, then it is our duty to ascertain on a contemporary scale what these objectives are."

Once upon a time, a person's beliefs were their private business. Now they are a United Nations edict. Muller continues, arguing that because humans have the capacity to think and feel, "to dream, to teach and to invent, the universe gives us an indication of what is expected of us: It wants us to know and to understand the maximum range possible of what the universe is all about.

"We are driven to know more and more of our globe and of Creation, including the art of recombining cosmic forces through energy, matter and life itself."

Notice there how Muller was not ultimately appealing to science, but to an "art"? Specifically, magic – "recombining cosmic forces". Really? From the United Nations? Sadly, yes.

Muller even equates the United Nations with God, daring to rewrite the verses from Genesis in the Bible:[279]

"And God saw that all nations of the earth, black and white, rich and poor, from North or South, from East and West, and of all creeds were sending their emissaries to a tall glass house on the shores of the River of the Rising Sun, on the Island of Manhattan, to stand together, to think together, and to care together for the world and all its people. And God said: "That is good". And it was the first day of the New Age of the Earth'."

Then, he says, become God yourself:[280]

"Decide to open yourself to God, to the Universe, to all your brethren and sisters, to your inner self...to the potential of the human race, to the infinity of your inner self, and you will become the universe...you will become infinity, and you will be at long last your real, divine, stupendous self ".

Muller was a lawyer by training, so where did he get his educational curriculum ideas from?

In the beginning of the Curriculum document, Muller says his ideas are "based on the teachings set forth in the books of Alice A. Bailey by the Tibetan teacher Djwhal Khul and the teachings of M. Morya."

This all sounds fairly regular, if perhaps a touch 'Dalai Lama-esqe', until you do the research and discover that Djwhal Khul and M. Morya were invisible two thousand year old spirit entities who allegedly communicated with Alice Bailey "telepathically".

Great, you may now be saying. 'My child's education is underpinned by the ravings of a woman who claimed to speak to dead people.' Yep, that about covers it. But Bailey wasn't just any raving seer. This next bit should give you a bit of a clue: she established the Lucifer Publishing Company in New York, and published a journal entitled *Lucifer*. She

279 "The Desire to be Human: A Global Reconnaissance of Human Perspectives in an Age of Transformation", ed. By Robert Muller, Miranana, 1983, p.17
280 "Decide to Be," by Robert Muller, Link-Up, 1986, p.2

was, in fact, a Satanist[281] and the high priestess of an occult religious movement known as "theosophy", used in some Montessori[282] and Rudolf Steiner[283] teaching. When she realised her Lucifer Publishing Company name was putting people off, Bailey renamed it the Lucis Publishing Company. Slightly different skin, same Beast. She also established, as a sister entity, the Lucis Trust. Her husband Foster Bailey was a 33rd degree Scottish Rite freemason.

Which brings us to those questions we posed a couple of pages back. Who is "the Absolute Supreme", the "God of all" that these United Nations top brass keep referring to?

---

281 Bailey and her followers would object to being called Satanists, because they argue Lucifer was seriously misunderstood and not at all the bad guy. Instead, he is an Angel of Enlightenment, sent to Earth to help make all men 'gods'. Nonetheless, 'Lucifer' originated in Judeo-Christian theology as Satan/Devil/Beelzebub/Mephistopheles/Morning Star. Their own documents clearly identify the object of their worship as "Satan". It is what it is. See http://www.lucistrust.org/en/arcane_school/talks_and_articles/the_esoteric_meaning_of_lucifer
282 "Montessori and the Theosophical Society" By Winifred Wylie, Quest magazine, Mar/Apr 2008, 53-55, http://www.theosophical.org/publications/1409
283 "Theosophy" by Rudolf Steiner, http://steinerbooks.org/research/archive/theosophy/theosophy.pdf

CHAPTER 19

Thought Control & The New Religion

"The choice for mankind lies between freedom and happiness and for the great bulk of mankind, happiness is better."
– George Orwell, 1984

Who is the stone altar in the heart of the UN building dedicated to? Here's a hint: the Lucis Trust is the Official Custodian of the Meditation Room for the United Nations, and has been for decades. The altar is dedicated to Lucifer – he who "is worshipped by many names". None other than what would be referred to in popular literature as "the Devil himself". You have to give the Devil his due, he has an astounding sense of theatre. There is no Christian chapel inside the UN building, but there is an altar to Lucifer.

Assistant Secretary-General Robert Muller was one of Bailey's disciples, and when he rose to prominence within the United Nations, co-founding UNESCO and ultimately picking up the global education prize in 1989, he arranged for Lucis to have official status at the UN, as a consultative organisation "on the roster" of the Economic and Social Council.[284] What does that mean? The UN website explains:[285]

---

284 See "List of non-governmental organizations in consultative status with the Economic and Social Council as of 18 September 2008", United Nations document, http://www.un.org/esa/coordination/ngo/pdf/INF_List.pdf
285 http://esango.un.org/paperless/Web?page=static&content=faqs

"Non-governmental, non-profit public or voluntary organizations may be admitted into a mutually beneficial working relationship with the United Nations by attaining consultative status with the Economic and Social Council (ECOSOC)."

Under that status, the Lucis Trust can "make a contribution to the work programmes and goals of the United Nations by serving as technical experts, advisers and consultants to governments and Secretariat."

"Sometimes, as advocacy groups, they espouse UN themes, implementing plans of action, programmes and declarations adopted by the United Nations. In concrete terms this entails their participation in ECOSOC and its various subsidiary bodies through attendance at these meetings, and also through oral interventions and written statements on agenda items of those bodies. In addition, organizations, qualifying for General Category consultative status, may propose new items for consideration by the ECOSOC. Organizations granted status are also invited to attend international conferences called by the U.N., General Assembly special sessions, and other intergovernmental bodies."

In other words, the Lucis Trust (sister organisation to what was the Lucifer Publishing Company) has been a trusted UN advisor for decades. Another Lucis subsidiary, World Goodwill[286], was based inside the UN headquarters for years, and a long time Lucifer disciple, expatriate New Zealander Steve Nation, has served as co-convenor of the Spiritual Caucus of the United Nations.[287]

But it's not just Lucis Trust and its subsidiary World Goodwill at work deep inside the UN operation. Another group with consultant status as an official NGO is the "Temple of Understanding", set up by another follower of the Lucifer doctrine, Juliet Hollister, in 1960:[288]

"In 1959, Mrs. J. Dickerman Hollister thought of the idea of a Temple of Understanding (the term was suggested by Mrs. Ellsworth Bunker, who was the wife of the American ambassador to India). Mrs. Hollister received encouragement from The Ford Foundation that same year, and the next year the syncretistic temple was founded by Juliet Hollister, with

286 World Goodwill's purpose is "to cooperate in the world of preparation for the reappearance of the Christ" One Earth magazine, Findhorn Foundation, October/November 1986, Vol. 6, Issue 6, p.24
287 http://www.spiritualcaucusun.org/
288 The Temple of Understanding also received initial funding from the Lucis Trust, John D. Rockefeller and Robert McNamara, who later became the US Defense Secretary. See "The Temple of Understanding" by Edith Kermit Roosevelt, The New Hampshire Sunday Times News, 21 October 1962

partial financing by the Carnegie Endowment for International Peace. The temple listed its "Founding Friends" as including Pope John XXIII, Thomas Merton, U. Thant, the Dalai Lama, and Eleanor Roosevelt. Listed among its advisory council members were Father Thomas Berry, Dr Robert Muller, and Brother David Steindl-Rast."[289]

Interesting to see one of its founders was the Pope at the time, along with Eleanor Roosevelt. Another name, not included in that quote, was Henry Luce III, publisher of *Time* magazine.[290] Luce and his publishing company endorsed the New Age, and sponsored theologians like James McCord who famously urged the Church to preserve its power by changing its beliefs: "If you ever have to choose between heresy and schism [breaking away], choose heresy."[291] Again, it's a testament to the old saying, power corrupts, and so does the fear of losing power. Others might call it being 'willing to do a deal with the Devil'.

It was Henry Luce's *Time* magazine that ran the provocative front cover in the late 1960s, "Is God Dead?"

Some of you reading this who were alive at the time will remember it, and remember the controversy that erupted. Luce's protégé James McCord claimed the death of God would usher in a "new era in theology".[292] Readers may have assumed it was objective journalism, few knew that Luce was tied up, ironically given his name, with the Lucifer movement and therefore had a vested interest in promoting the Death of God. Luce would not have helped set up the "Temple of Understanding" and funded heretic priest James McCord, if he did not have a dog in the ring. These people played a long game.

You may be asking why the leading branch of Satanism is so entwined with the United Nations that they are running its religious services? Good question. It all goes back to the writings of Alice Bailey who predicted Lucifer (referred to in Theosophy and New Age writings deliberately and misleadingly as '*the* Christ'[293]) could only reappear if enough people began to believe in him and ask him to come:

---

289 "Now is the Dawning of the New Age New World Order" Dr. Dennis Cuddy, Hearthstone, 2000 p.143
290 http://templeofunderstanding.org/who-we-are/founding-friends/
291 http://www.fpc-bethlehem.org/_PDF/Ten-Theological-Challenges-Facing-the-PCUSA.pdf
292 "Toward a Hidden God: Is God Dead?", *Time* 8 April 1966, p82
293 The term 'Christ' is a title, not a name. In Greek it meant the Anointed One, and signified the Messiah. In Luciferian belief, Jesus was a usurper and the true 'anointed one' is Satan/Lucifer, the force behind the Anti-Christ.

'The Hierarchy is at this time attempting to channel the forces of reconstruction into the Assembly of the United Nations. (The forces of reconstruction are related to the Will aspect of divinity and are effective mainly in relation to those *entities,* which we call nations.) The use of these impersonal energies is dependent on the quality and the nature of the recipient nation; on its measure of true enlightenment and on its point in evolution. Nations are the expression today of the massed self-centredness of a people and of their instinct to self-preservation. The main object of the Hierarchy is so to distribute these constructive energies that the theory of unity may slowly be turned into practice and the word "United" may come to have a true significance and meaning.' Also in The Reappearance of the Christ it is stated that *the one who works to produce at-one-ment, unification and fusion is generating a slowly growing will-to-unity within the Assembly of the United Nations. This being can only channel His energies through the mass consciousness or through a group conscious entity, such as the U.N."* [emphasis added]

Now, if you're like most people you'll probably view the above as a load of old bat guano, which is totally understandable; it would also be a dangerous underestimation of the seriousness of the situation. It doesn't actually matter what *you* think of the spiritual beliefs infiltrating the UN bureaucracy, because *you* won't actually be in charge of the global governance framework when it comes – *they will.*

Let me put it another way. How comfortable would you feel if radical Islamists had wheedled themselves into positions of enormous power and influence in the United Nations on the eve of the imposition of a global governance system? Or what if the UN was run by born again Christians intent on converting the entire globe through legislation and re-education?

If you don't feel comfortable with either of those scenarios, nor should you be comfortable that Satanists genuinely believe they can usher in the emergence of the Anti-Christ through the mechanism of the United Nations and the policies they introduce. And yet, look at what they have achieved, preaching the message of world peace through the unity of humanity, under global law. While recognising the seductive nature of their message to secular ears (who could object to "world peace" or "unity"?), it appears from Bailey's writings quoted above that they see those mantras primarily as consciousness-raising and a spiritual requirement to bring in their 'Master'. They believe that if they can get enough people in the world behind their particular brand of peace activism, it will act as a tipping point.

The Christian Bible did warn the devil would come disguised as an "Angel of Light", and you couldn't get much lighter than the lofty promises of peace and brotherhood. The very name 'Lucifer' means light-bearer.

On 20 May 1999, two "Lightworkers" named Steve and Barbara Rother held a spiritual channelling session for 70 United Nations staff at the UN complex in Vienna, Austria. This site houses, among a raft of other units, the Atomic Energy Agency.

In a lengthy report on their séance, Steve Rother wrote, "I was surprised to find that most of the people in the room had been reading the Beacons of Light Meditations for some time. It seems that the Messages from the Group get passed around in the internal email system of the UN. They were ready for the information and very anxious to see us."[294]

If you are wondering, "the Group" is a reference to some of the spiritual Hierarchy Alice Bailey wrote about, the demonic entities supposedly guiding the return of "the Coming One". Steve Rother claims to hear messages from "the Group" to be relayed to earthlings.

Rother says the UN staff wanted to hear from "the Group" about why the world was going to hell in a handcart – the siege of Kosovo nearby was topical. "The Group" told the UN staff that they and their organisation were to play a pivotal role in Earth's future, "and how their sacrifice would facilitate the further advancement of humanity." The Group said that humanity's "awakening on this planet began in earnest the day that Hitler left. It was our collective raise in vibration that said we would never give our power away like that again."

The "Group" – through channeller Rother – instructed the UN staff in the room that all of humanity was to be blended together and that "those actions that resisted blending...would not be supported".

The spirit entities then paraphrased the well-worn sci-fi cliché: 'resistance is futile'. The new spiritual energy and vibrations sweeping the planet as believers adopt New Age and Buddhist beliefs was crucial, they warned, and if ever people stopped believing and "misdirected" their energies against the Coming One and the plan, "then the Mother would end the Game and the cataclysmic end times would be visited upon the Earth."

---

294 The Lightworker website, now only accessible through the Internet Archive. It features photos of the large audience. http://web.archive.org/web/20000823093443/http://www.lightworker.com/articles/unitednations/un052099.shtml

The "Group" also told the UN staff they had to be prepared to fight wars in the name of the Coming One:

"They made it clear that we were in the higher vibrations and these actions would now meet abrupt opposition. They said that the words 'fight' and 'war' would not be in our vocabulary long, but it was also very important that we took a stand...they greatly honoured those who accepted the role of victim."

Not all peace and love then.

To get a better feel for how seriously the United Nations takes its Satanism, consider this: the Meditation Room in UN HQ New York, with the block of rock and the light beam, is directly under the General Assembly chamber, in a very privileged spiritual position.[295] Its custodian is the Lucis Trust. It is the only religious room in the UN. Its followers channel "spirits" to guide the UN representatives in the room above during key debates:

"When I was introduced to the Aquarian Age Community[296] at the United Nations," writes New Ager Susan MacNeil, "I found the service activities, especially the meditation outline, deeply inspiring. As I began to participate in this group service with the UN meditation, my urge to serve increased. At the same time, I was observing a number of Ambassadors and other UN staff members being introduced to the UN meditation through Aquarian Age workshops and connecting to the ideas presented in the meditation outline."[297]

In other words, the Lucis Trust is ensuring as many key UN players as possible join their belief system.

"With the UN meditation we are uniting the planetary heart centre (the Hierarchy), with the planetary head centre (Shamballa), in order to access the Plan that is an expression of Purpose and produces intelligent activity within the human kingdom. I believe the meditation is having a powerful impact on the UN and in society that is observable in specific ways; for example, Secretaries-General, Kofi Annan and Ban Ki-Moon have both referred to the importance and significance of the UN Meditation Room," says MacNeil, who attributes the 'Lucifer' prayers to creating "greater demands from the leaders of all countries for right human relations towards the common good, the social media connecting

---

295 Although it is currently closed for two years while the UN building undergoes a $2 billion refurbishment

296 Another Lucis Trust offshoot

297 "The Spiritual Work Of The United Nations And The Liberation Of Humanity", By Susan MacNeil, Ph.D., Newsletter of the Aquarian Age Community 2012 No. 2, http://www.aquaac.org/dl/12nl2art3.html

TOP LEFT: Theosophy founder Helena Blavatsky. Regardless of what New Age devotees tell themselves about the meaning of Lucifer, her Secret Doctrine was very clear: "Satan is...worshipped by many names...the god of this planet". TOP RIGHT: The logo used by the United Lodges of Theosophy. Serpent, swastika, pentacle, ankh. BELOW: The General Assembly Hall of the United Nations. Directly below the hall is the altar.

like minds everywhere, and uprisings by citizens of the world expressing their will through various actions.

"Learning that the meditation room is housed right below the General Assembly seemed to signify another significant alignment, and after many years of using the meditation, I experience the energy, strength, and force behind its use. It is through visualization, imagination and the will that we engage the awakening of human minds and hearts toward our common thread of Oneness on this planet. What other body can best accomplish this other than the UN? I sometimes have the benefit of meditating in the UN's General Assembly, and use the meditation by first visualizing the Avatar of Synthesis, the Spirit of Peace, and the Lord Buddha overshadowing[298] the Christ, standing as the head of the General Assembly, with the intention of strengthening the hands of the New Group of World Servers."

Surely, you mutter, this is just mumbo-jumbo, a bunch of airy-fairy types wearing tinfoil hats jumping around in a circle underneath the debating chamber whilst flinging feathers and gold dust into the air and chanting something suitably exotically Eastern; file it under "mostly harmless".

Again, you are missing the point. It's not the "efficacy" of the Luciferian prayer rituals that is at issue, it is the fact that the Lucifer disciples long ago captured the hearts and minds of UN policy bureaucrats, thus ensuring that their belief system is entrenched in UN policies worldwide.

You are probably also seriously underestimating just how widespread this belief system now is. The Lucifer doctrine derives from Theosophy, whose two major writers were Helena Blavatsky and Alice Bailey. Contrary to popular belief, most modern Buddhism in the West, particularly the Zen strain, is not authentic Buddhism but a hybrid largely influenced by Theosophy.[299] In other words, modern Buddhism is the occult in drag, made palatable by its seeming exoticness.

---

298 An occult phrase akin to "possessing"
299 "The Making of Buddhist Modernism" by David L. McMahan, Oxford University Press, 2008, see also "The Divinity Code" by Ian Wishart, Howling At The Moon Publishing, 2007. In 1995, scholar Robert Sharf wrote, quite pointedly: "The popular 'lay image of Zen, notably the notion that Zen refers not to a specific school of Buddhism but rather to a mystical or spiritual gnosis that transcends sectarian boundaries, is largely a twentieth-century construct. Beginning with the persecution of Buddhism in the early Meiji (haibutsu kishaku) Zen apologists have been forced to respond to secular and empiricist critiques of religion in general, and to Japanese nativist critiques of Buddhism as a 'foreign funerary cult' in particular. In response, partisans of Zen drew upon Western philosophical and theological strategies in their attempt to adapt their faith to the modern age." See http://www.thezensite.com/ZenEssays/CriticalZen/whose%20zen_sharf.pdf

The criticism has often been made that climate change policy is almost religious in the way it is being rolled out. It is, and that's because it is central to the Gaia Earth-worship and Lucifer "consciousness-raising" that the UN is trying to engender. Around half the planet "believe" in climate change, almost in the religious sense, including most of the mainstream media. They "believe" it because it is being pushed as a doctrine of faith, disguised as science, by the Lucis Trust and its followers worldwide. If you want proof of their power to persuade, and therefore the danger inherent in their ideas, it is staring you in the face. They are "mostly harmless" only in the same sense that a cyanide pill is.

When Global Governance comes in the next decade as they are planning, it will come with this baggage.

In 2007 the *Canada Free Press* published an opinion piece, noting that the new UN Secretary General Ban Ki-moon's very first act, after being elected to the top position that week, was to visit the Lucifer chapel:

"The bizarre brainchild of the late Dag Hammarskjöld, the UN Meditation Room is built in the shape of a truncated pyramid," reported Judy McLeod in the *Free Press*. "In the center is an altar made out of magnetite, the largest natural piece of magnetite ever mined. For meditation purposes it is probably the most ideal spot on the planet, since the magnetite altar has its foundation straight down, built into the bedrock of the land below; tapping into the energies of the earth itself. The mysterious mural also helps the worshippers tune into esoteric energies, and helps facilitate a state of altered consciousness.

"No self-respecting Christian would come to pray at the Meditation Room whose custodian is the Lucis Trust, formerly known as the Lucifer Publishing Co. The room was a favourite haunt of former UN Secretary-General Kofi Annan, who was married to his Swedish wife there.

"That's where Ban headed right after the welcoming applause accorded him by dozens of UN staffers."[300]

Odd, if it is just mumbo-jumbo, that Secretaries-General of the UN bow their knees regularly at an altar dedicated to Lucifer.

Let's remind ourselves again just who this Lucifer character is. Clearly the UN bigwigs and some of their senior bureaucrats and ambassadors are followers, so what do we know from the various strains of religious history?

"How art thou fallen from heaven, O Lucifer, son of the morning!," says

---

300 "Ban Ki-Moon & Kofi Annan, United Nations Bobbsey Twins" by Judi McLeod, Canada Free Press, 4 January 2007, http://www.canadafreepress.com/2007/cover010407.htm

the Old Testament of the King James Bible. In the New International Version of the Bible the name 'Lucifer' is missing: "How you have fallen from heaven, O morning star, son of the dawn!"[301]

Another verse a few moments later bears some eerie resemblance to the Lucifer temple underneath the United Nations General Assembly room: "You said in your heart, I will ascend to heaven, I will raise my throne above the stars of God; I will sit enthroned on the mount of assembly."

As the biblical version goes, Lucifer led a revolt in heaven and was cast out, taking with him a host of fallen angels (now referred to as 'demons' to distinguish them from the holy angelic) to manifest themselves in the physical universe. The Bible more commonly refers to this being as Satan, such as the comment of Jesus Christ in the Gospel of Luke: "I saw Satan fall like lightning from heaven."[302]

In the New Testament, the apostle Paul writes: "Satan himself masquerades as an angel of light. It is not surprising then if his servants masquerade as servants of righteousness. Their end will be what their actions deserve."[303]

Light-workers, take note.

In the Koran, Satan is known as 'Shaytan', and he even makes an appearance in the Hindu vedic scriptures as the death god Kali who confronts Easa Maseeha (Jesus the Messiah).[304]

The Lucifer prayed to by the UN officials is defined for them by Theosophy: the teachings of Madame Helena Blavatsky as refined by Alice Bailey. As previously noted, Theosophy sees Lucifer as "the light of the world" and the real name behind the various names humans have for God. There's a little bit of bait and switch going on with that game, however, because in most major Western religions (Christianity, Islam, Judaism) the central figure of worship is God the Creator, not one of the angelic subset.

Blavatsky's book *The Secret Doctrine* spells out Theosophy's direct worship of Satan himself as "saviour":

"Lucifer represents.. Life.. Thought.. Progress.. Civilization.. Liberty.. Independence.. Lucifer is the Logos.. the Serpent, the Saviour."[305]

---

301 Isaiah 14: 12-15
302 Luke 10:18
303 2 Cor. 11:14-15
304 Bhavashya purana- Prathisarga parva, IIIrd part- 2ndchapter- 23rd verse
305 The Secret Doctrine, by Helena Blavatsky, pages 171, 225, 255 (Volume II)

"It is Satan who is the god of our planet and the only god."[306]
"The Celestial Virgin which thus becomes the Mother of Gods and Devils at one and the same time; for she is the ever-loving beneficent Deity...but in antiquity and reality Lucifer or Luciferius is the name. Lucifer is divine and terrestrial Light, 'the Holy Ghost' and 'Satan' at one and the same time."[307]

Everything reduces to One in Theosophy, and the One is Lucifer, and Lucifer is Satan, and "Satan is the god of our planet and the only god."

Let's make this absolutely clear: if you are a tofu and mung-beans kind of person who dabbles in a bit of Buddhism or a bit of New Age self-improvement, it is time you understood who the people that founded the New Age movement and devised the meditation strategies really say you are tapping in to.

As the Genesis story goes, Lucifer was the Serpent in the Garden of Eden who tempted Eve from the Tree of Knowledge. Most of you are automatically thinking "apple", and the temptation, you might assume, was fruit. But you'd be wrong. The fruit was merely the key to the promised kingdom. What was the promised kingdom Satan offered?:

"In the day ye eat thereof, then your eyes shall be opened, and ye shall be as gods." (Gen. 3:5 KJV)

Now, let's take a closer look at Robert Muller's words on the world school curriculum:[308]

"As it is vividly described in the story of the Tree of Knowledge, *having decided to become like God* through knowledge...*we have also become masters in deciding between good and evil.*"

In other words, "we shall be as gods".

Muller urged that educators use this opportunity of a core curriculum to "teach children and people a sense of participation and responsibility in the building and management of the Earth, of becoming artisans of the will of God and of our further human ascent. A new world morality and world ethics will thus evolve, and teachers will be able to prepare responsible citizens, workers..."

Muller talked of "human ascent", and this is the New Age belief in progress to a higher level of consciousness and being, again – the idea

---

306 Ibid, pages 215, 216, 220, 245, 255, 533, (VI)
307 Ibid, p539
308 "The Need for Global Education" by Robert Muller, New Era magazine, World Education Fellowship, January 1982

that we "are as gods". We will become, Muller says, "what we were always meant to be: universal, total beings. The time for this vast synthesis, for a new encyclopedia of all our knowledge and the formulation of the agenda for our cosmic future has struck."

If you remember the lesson from Genesis, the Serpent was booted out of heaven for the arrogance of pride in his achievements. Here's what Muller wants children to be educated to become:

"Education…is to make each child feel like a king or queen in the universe, an expanded being aggrandized by the vastness of our knowledge. It is to make each human being feel proud to be a member of a transformed people.

"A world core curriculum might seem utopian today," Muller wrote in 1982, "[but] by the end of the year 2000 it will be a down-to-earth, daily reality in all of the schools of the world."

Bat guano you might think, but Lucifer disciple and UN Assistant Secretary-General Robert Muller's educational theories based on Satanism and earth religion are now taught in virtually every major school curriculum in the West. The importance of the United Nations, for example, is a dominant social studies curriculum item in New Zealand, as is the mantra of "global solutions for global problems". The Gaia theory runs rife in science and social studies classes. Tolerance of all religions, and banning students from making truth claims about a religious belief are part of the same educational package being rolled out. Religion is not being 'banned' from schools, it is being redefined as holistic and all encompassing. Many of Muller's concepts are woven into the new "Common Core" spreading through American schools.

As Muller wrote: "How can our children go to school and learn so much detail about the past, the geography and the administration of their countries, and so little about the world, its global problems, its interdependencies, its future and its international institutions? People are astonished by the sudden emergence of global crises. They wonder how environmental degradation could have developed to the point of endangering life on this planet.[309]

"It is therefore the duty and the self-enlightened interest of governments to educate their children properly about the type of world in which they are going to live. They must inform the children of the actions, the

---

309 "The Need For Global Education", by Robert Muller, reprinted http://robertmuller.org/pdf/RobertMuller-GlobalEducation.pdf

endeavours and the recommendations of their global organisations. "Institutes for global education have sprung up, and the UN and UNESCO are convening meetings of educators to develop global curricula...the world will be in great trouble and will not be able to solve its global problems if citizens are not taught properly from their earliest youth." Muller advocated the creation of a one world religion/ethical system capable of uniting the planet, and said it was the seeds of this new philosophy that had to be ever so subtly planted in schoolchildren. They had to be taught that claims of religious exclusivity, such as the words of Jesus – "I am the way, the truth and the life, no one comes to the Father but through me" – were wrong, non-inclusive and hateful, that all religious paths were equally valid ways of connecting to God.

And you thought your kids were going to school to learn the basics? Read between the lines of the lessons a little more carefully. Muller wrote of developing "new codes of behaviour which will encompass all races, nations, religions and ideologies. It is the formulation of these new ethics which will be the great challenge for the new generation. It will concern not only man's material fate but also his mental and spiritual lives.

"Yes, global education must transcend material, scientific and intellectual achievements and reach deliberately into the moral and spiritual spheres.

"Global education must prepare our children for the coming of an interdependent, safe, prosperous, friendly, loving, happy planetary age as has been heralded by all great prophets. The real, the great period of human fulfilment on planet Earth is only now about to begin."

Then there's the strange case of Lumbini, in mountainous Nepal. Virtually every UN Secretary-General in history has made a pilgrimage to Lumbini, a remote archaeological site 10 hours' drive from the Nepalese capital, Kathmandu, in the foothills of the Himalayas. Most readers will be unaware that Lumbini is the reputed birthplace of Buddha.[310]

Dag Hammarskjöld visited Lumbini in March 1959, and paid homage in a haiku verse:

"Like glittering sunbeams/The flute notes reach the gods/In the birth grotto."[311]

U Thant made the UN leader's pilgrimage in 1967 and called it "one

---

310 Most scholars believe it was a different site, but the UN has honed in on Lumbini. The site is sacred to the Theravada strain of Buddhism which is directly linked to the Lucis Trust, as you will discover later in this book.
311 "United Nations Secretaries-General and Lumbini", UNESCO document, http://www.unesco.org/new/index.php?id=66087

of the most important days in my life", when he wrote his memoir in 1977, *View From The UN.*

Kurt Waldheim's[312] tour in 1981 resulted in an announcement that the UN would spend donations on a "Lumbini Master Plan" to regenerate the ruins as a spiritual focal point for humanity.

"Through the efforts of the Government of Nepal and with financial assistance from the United Nations Development Program, a Master Plan has been completed by the Japanese architect Kenzo Tange. However, a necessary is to make these plans a reality. It is my hope, therefore, that government, private institutions and individuals will make generous contributions towards this most worthy undertaking."

In 1989 the next Secretary-General, Javier Perez de Cuellar, spoke in glowing terms of Lumbini:

"For all mankind Lumbini has special meaning as a place of meditation and spiritual renewal, a center of culture exchange and a symbol of peace. Buddha's message of compassion and devotion to the service of humanity is more relevant today than at any other time in history."

De Cuellar added that the UN saw the site playing a key role in "the spiritual and cultural heritage of humanity."

Kofi Annan, in 1998, spoke also of how "Lumbini provides yet another illustration of the inter-connectedness of all people, across borders and across time. As a United Nations Educational, Scientific and Cultural Organization (UNESCO) World Heritage site, Lumbini reminds us how much the world's religions can teach us, Buddhists and non-Buddhists, believers and non-believers alike."

The current Secretary-General Ban Ki-moon visited 2008 and said:

"I am awestruck by the beauty and profound significance of this site, the birthplace of the Lord Buddha. Being here, I am reminded of his amazing life journey from sheltered prince to founder of one of the world's great religions. And I am moved by his example of voluntarily leaving behind comfortable circumstances to confront the painful realities of life and to help others overcome them. Above all, as Secretary-General of the United Nations, I am all the more inspired to work for peace throughout the world. I sincerely hope that we can learn from his lessons, from his teach-

---

312 Waldheim had been suspected by a UN War Crimes Tribunal in 1948 of being a Nazi war criminal but they did not have sufficient evidence. Incredibly he became UN Secretary-General despite those suspicions. In 1986 the truth about Waldheim's past came out. "Kurt Waldheim dies at 88; ex-UN chief hid Nazi past", NY Times, 14 June 2007, http://www.nytimes.com/2007/06/14/world/europe/14iht-waldheim.3.6141106.html?_r=0

ings and his philosophy to bring peace, stability, harmony, reconciliation and friendship among people of different beliefs, different religions and cultures. This is exactly what human beings should promote and pursue for a better world, a more peaceful, more prosperous world."

Sharp-eyed readers will have spotted Ban Ki-moon's reference to "Lord Buddha". You won't find a UN speech where Ban refers to "Lord Jesus". The UN's commitment to religious neutrality only goes so far.[313]

The United Nations agency co-founded by Robert Muller, UNESCO, has listed Lumbini as a world heritage site. The UNESCO page also refers to the site not in neutral archaeological terms but spiritual ones, again giving Buddha a spiritual honorific:

"Lumbini, the Birthplace of the Lord Buddha."

In the UN list of World Heritage sites, each has a number. Wait till you see what the number of this particular world heritage site is:

---

313 A quick search of the UNESCO site lists some 11,000 documents with the phrase "Lord Buddha", and 13 with the phrase "Lord Jesus" – none of the latter were issued by UNESCO directly, unlike its constant press release references to "Lord Buddha".

CHAPTER 20

One Ring To Rule Them All, And In The Darkness Bind Them

"That name the hobbits only knew in legends of the
dark past, like a shadow in the background of their
memories; but it was ominous and disquieting"
– J R R Tolkien, LOTR

In the Book of Revelation in the Bible, a much-talked about passage talks about the end of times and the emergence of something known as "the Beast". The number of the Beast "is 666".[314]

The number of Lumbini is 666. In fact, specifically, "666Rev".[315]

As previously noted, you couldn't make this stuff up.

The word "Buddha" is from ancient Sanskrit and literally means "the Awakened One", or "the Enlightened One".[316] It is not a name, but a title. In substance, the Buddha is an Eastern version of the Lucifer doctrine. Buddha is the East's light-bearing Lucifer, whose name also bears awakening and light connotations: "morning star", "son of the dawn" or "light-bearer". Is it a coincidence that these two ancient names, in their own languages, both mean essentially the same thing? Especially significant, perhaps, that the Buddha's birthplace is marked "666" in the United Nations system.

---

314 Rev 13:18
315 http://whc.unesco.org/en/list/666
316 http://www.yowangdu.com/tibetan-buddhism/buddhist-mantras-shakyamuni.html

Whatever your religious views, the preceding paragraphs should give you immense cause for concern. The whole concept of separating Church and State was to prevent a government from effectively dictating what you should believe. Yet here you have the parliament of governments, the United Nations, actively working to bring in a new religious belief system to unite the planet.

Speaking in 2000, UN Secretary-General Kofi Annan again acknowledged that spiritualism drove the United Nations from its core:

"For many of us, the axiom could well be. 'We pray, therefore, we are'."

Then, borrowing a quote from Martin Luther King, Annan added, "This says to us that our world is geographically one. Now, we are faced with making it spiritually one. Through our scientific genius we have made of the world a neighborhood; now through moral and spiritual genius, we must make it a brotherhood."

Why is it the job of the United Nations to make all humanity "spiritually one"?

You might be thinking, well, Robert Muller died in 2010, Alice Bailey is long dead, surely no one is still driving this agenda at the UN? Think again. Remember Kofi Annan got married in the Lucifer meditation room, and that altar was the first stop for Ban Ki-moon after he was elected Secretary-General in 2007.

If you want to know just how deep all this goes, consider this: The Lucis Trust has published a 'prayer' purportedly given to Alice Bailey in 1945 at the same time as the UN was being established, by the 2000 year old spirit creature Djwhal Khul. It is called "the Great Invocation" and it is a prayer to the entity Lucifer/Satan/Maitreya/Whatever to return to Earth as ruler.

At one point a few years back, the Lucis Trust was slightly more open about its plans to institute a replacement world religion through the UN, posting this Alice Bailey morsel on its website:

"The Great Invocation if given widespread distribution, can be to the new world religion what the Lord's Prayer has been to Christianity and the 23rd Psalm has been to the spiritually minded Jew."

That paragraph can now only be located through the Internet Archive system. The "invocation", or spell, has certainly gained wide distribution. It was the opening 'prayer' for the pre-conference of the 1992 Earth Summit in Rio – the one that ramped up the climate change steamroller. The Great Invocation has even been endorsed by the UN as a prayer for International Peace Day:

"Please add the radiations of your mind and heart to a Global Vigil of Invocation, Meditation and Prayer in support of International Day of Peace, 2013. To support this Day on a spiritual level, there will this year be a wide variety of initiatives for silence, meditation and prayer. The UN continues its important focus, inviting the people of the world to observe a minute of silence at 12 noon local time, setting up a wave of silence beginning in New Zealand and flowing around the world to end in Samoa almost 48 hours later. A range of movements are collaborating for the first time with the vision of co-creating the Largest Globally-Synchronized Meditation and Prayer for Peace in human history.

"Will you or your group commit yourself to … beginning and ending the period with a sounding of the Great Invocation, or the Peace Invocation 'May Peace Prevail on Earth', or an invocation or prayer for world peace of your choice? Imagine the rhythmic pulse of invocation flowing from around the globe every 15 minutes."

What exactly is the Great Invocation? Whilst its proponents would never say as much, it is a prayer inviting the emergence of the Anti-Christ:

"Let light stream forth into human minds./Let Light descend on Earth…./May the Coming One return to Earth."

Who is "the Coming One"? The clue is the "light". He's Lucifer, the morning star, the son of dawn, the serpent – as Blavatsky wrote. He is why a six and a half tonne block of iron ore is bathed in light from an unseen source above, immediately beneath the UN debating chamber, an altar, as the UN Secretary-General called it "to the God of all".

You couldn't write a movie with a script like this, no Hollywood producer would believe it. Yet there it is: the United Nations, the entity that desperately wants our politicians to sign global treaties for a global governance structure, holds within it, in pride of place, an altar to Satan. You'll recall the UN Secretary-General was a follower of Alice Bailey and Helena Blavatsky's Luciferianism, and that Blavatsky wrote "It is Satan who is the god of our planet and the only god."[317]

By a process of elimination, there is no other entity the UN temple could be dedicated to.

The climate change Earth Summit at Rio in 1992 was kicked off in its preliminary session, a Sacred Gathering, with a prayer to Satan. It was the largest gathering of world leaders in history. Little wonder the climate debate has such religious overtones. An attempt to stir up the biggest changes in human consciousness in history is being done in the name of Lucifer. You could be forgiven for thinking the world truly is going to Hell in a handcart.

Again, and this point cannot be overstressed: regardless of whether *you* believe any of the supernatural jargon, the people ultimately controlling this and driving the UN process do, and so far they're winning.

The Lucis Trust, incidentally, mocks ordinary members of the public and politicians who don't truly know what they're saying when they repeat the Great Invocation spell. In an Alice Bailey publication several years ago, they said the Invocation would be understood in three different ways: 1. cattle class, 2. Esoteric and initiates, 3. By the Masters of the Hierarchy:

"The general public will regard it as a prayer to God Transcendent. They will not recognise Him yet as immanent in His creation; they will send it forth on the wings of hope – hope for light and love and peace, for which they ceaselessly long. They will also regard it as a prayer for the enlightenment of all rulers and leaders in all groups who are handling world matters; as a prayer for the inflow of love and understanding among men, so that they may live in peace with one another; as a demand for the working out of the will of God – a will of which they can know nothing and which ever seems to them so inscrutable and so all-inclusive that their normal reaction is patience and a willingness to refrain from questioning; as a prayer for the strengthening of human responsibility in order that the recognised evils of today – which so distress and trouble mankind – may be done away with and some vague source of evil may

---

317 The Secret Doctrine by Helena Blavatsky, Vol II

be harnessed. They will regard it finally as a prayer that some equally vague primeval condition of blissful happiness may be restored and all unhappiness and pain disappear from the earth. This is, for them, entirely good and helpful and all that is immediately possible."

Did you read the contempt for the great unwashed in that last line, that the public's shallow understanding of the real meaning of the prayer is "all that is immediately possible"? This is directly relevant to the line in the Invocation calling on the Coming One to guide the "little human wills" of men.

Now see how wise and clever the adepts and initiates of the occult are (as they claim) in regards to understanding the Invocation:

"Esotericists, aspirants and spiritually minded people will have a deeper and more understanding approach. To them it will convey the recognition of the world of causes and of Those Who stand subjectively behind world affairs, the spiritual Directors of our life."

The capitalisation in there is deliberate, and in the original text. The Lucis Trust is speaking of entities they regard as holy. In the official pantheon of the supernatural, you might better understand them by another name: demons. Think of *The Omen*, you'll get the picture, or the menacing Ring Wraiths in Tolkien's *Lord of the Rings*.

"They stand ready to strengthen those with true vision," writes the Lucis Trust, "ready to indicate not only the reason for events in the various departments of human living, but also to make those revelations which will enable humanity to move forward out of darkness into light. With this fundamental attitude, the necessity for a widespread expression of these underlying facts will be apparent and an era of spiritual propaganda, engineered by disciples and carried forward by esotericists, will mature.

"This era began in 1875 when the fact of the existence of the Masters of the Wisdom was proclaimed. It has been carried forward in spite of misrepresentation, attack upon the concept, and scorn. Recognition of the substantial nature of the available evidence and the appearance of an intuitive response by occult students and many of the intelligentsia throughout the world has been helpful."

Interesting to see the tip of the hat to how the "intelligentsia" swung in behind the spirit of the Invocation.

Finally, says the Lucis Trust, there are outliers, people with special insight into the supernatural quickening taking place:

"Thirdly, both of these groups – the general public and the world aspirants in their varying degrees – have among them those who stand out from the general average as possessing a deeper insight and understanding; they occupy a no-man's-land, intermediate on the one hand between the masses and the esotericists and, on the other, between the esotericists and the Members of the Hierarchy. Forget not, They also use this great Invocation and that not a day goes by that the Christ Himself does not sound it forth.

"The use of this Invocation or Prayer and the rising expectancy of the coming of the Christ hold out the greatest hope for man-kind today. Great Sons of God have ever come on humanity's demand and always will, and He for Whom all men wait today *is* on His way."

As you can see, all those teenagers going to peace meetings and climate change activism training are what the Lucis Trust effectively calls the "useful idiots" – people who are basically blind to the intention and powers that the occultists are trying to invoke but who are willing to go along with it in the name of peace, love and goodwill.

At the time of writing you could not read that briefing on the Lucis Trust website, it appears to have been removed or placed in a different directory. It is however accessible through the Internet Archive.[318]

A good analogy to this is the computer scam known as "phishing". That's where hackers masquerade as something they are not to fool you into cooperating with them. You can see in the paragraphs above the sneering contempt the Lucis Trust has for ordinary people; in fact, the "Great Invocation" spell calls for the 'Coming One' to dominate the people and their "little human wills". Yet the Lucis Trust has no qualms about masquerading as an organisation promoting peace, light, love, harmony and goodwill to all humans. What's not to love about that, on paper?

So millions sign up and have their "consciousness" raised by the self-proclaimed acolytes of Lucifer, blissfully unaware they are being used, in every sense of the word. They are used as cannon-fodder and bums on seats at protest rallies; they are used as sheep to be fleeced for financial and time donations to the various "peace" and "climate" causes; and for those who believe in the supernatural, as around 90% of the planet does on current polling, then by definition your supernatural energy is being "phished" and hijacked by a cult praying for the return of Satan. Is that truly what you signed up for?

---

318 http://web.archive.org/web/20060106071714/http://www.lucistrust.org/invocation/gi.shtml

Robert Muller wrote that the 'Coming One's return had been "heralded by all great prophets". Let's take a brief look in light of that old proverb, "be careful what you wish for":

"Many religions believe that a World Teacher will return to earth," writes New Zealand-born, American-resident Satanist Steve Nation[319], "knowing this Coming One by such names as Christ, The Lord Maitreya, the Imam Mahdi and the Messiah."

Let's look at the literature however: when this entity returns, the prophets predict global carnage: Armageddon. Yet on International Peace Day, that's the spell they were chanting, calling for this entity to show itself. It does seem very ironic for so-called peace activists to be wishing that on the world.

In the Islamic Ha'dith, al-Mahdi is a man who returns just before the Day of Judgement, rules Earth under Islamic sharia law for seven years, and dies.[320]

The Christian (and Jewish) messianic prophecies are detailed best in the Book of Daniel, where the prophet talks of "a ruler who will come… He will confirm a covenant with many for one 'seven'. In the middle of the 'seven' he will put an end to sacrifice and offering. And on a wing of the temple he will set up an abomination that causes desolation, until the end that is decreed is poured out on him."

In other words, at the end of a period of seven years, the 'Coming One' bites the dust. Not before, however, wreaking havoc. In the middle of the 'seven' this ruler moves decisively against the Church. Daniel identifies this as a ruler from an Iron kingdom – apt perhaps that the largest block of iron ore ever mined sits in the UN headquarters as an altar. The prophet writes that this Anti-Christ will "speak against the Most High and oppress his saints and try to change the set times and the laws."

The moves against the "set times" Christian festivals Christmas and Easter are rising up everywhere, seemingly in fulfilment of prophecy. In the New Testament, Jesus Christ warns that as the end of the world approaches, so too would an impostor rise up:

"At that time if anyone says to you, 'Look, here is the Messiah!' or 'Look, there he is!', do not believe it. For false Christs and false prophets will appear and perform signs and miracles to deceive the elect – if that were

---

319 Nation is a senior figure in the Lucis Trust network and its World Goodwill UN NGO offshoot
320 http://www.irshad.org/islam/prophecy/mahdi.htm

possible. So be on your guard; I have told you everything ahead of time."[321]

For three and a half years, says the Book of Daniel, this persecution is unleashed. It is only the return of Jesus Christ himself (not 'a' Christ, or 'the' Christ, but Jesus) that brings the whole shooting match to an end.

The Book of Revelation adds to the theatre. It talks of a "false prophet" who fools "the kings of the earth" into swearing allegiance to the Anti-Christ. "Let him that hath understanding count the number of the beast: for it is the number of a man; and his number is 666."[322]

This Anti-Christ figure apparently has supernatural powers, "so that he maketh fire come down from heaven on the earth in the sight of men… power was given him over all kindreds, and tongues, and nations. And all that dwell upon the earth shall worship him, whose names are not written in the book of life."

Revelation then talks of the second coming of Jesus Christ, and a final confrontation at "a place called in the Hebrew tongue Armageddon".

That place, the plain of Megiddo, is in modern Syria.

The kings of the earth surrendered their sovereignty "by agreeing to give the beast their power to rule", which, if global governance is rolled in by the United Nations, could effectively be ticked off the prophecy bucket-list as "done that".

"Then I saw the beast and the kings of the earth and their armies gathered together to make war against the rider on the [white] horse and his army. But the beast was captured, and with him the false prophet who had performed the miraculous signs on his behalf. With these signs he had deluded those who had received the mark of the beast and worshipped his image."

And from there, it was all over Rover. *It does not end well for the false 'light-bearer' and his 'light workers'.*

In the Book of Daniel, the Archangel Gabriel is quoted as saying: "When their sin is at its height, a fierce king, a master of intrigue, will rise to power. He will cause a shocking amount of destruction and succeed in everything he does. He will destroy powerful leaders and devastate the holy people. He will be a master of deception, defeating many by catching them off-guard. Without warning he will destroy them. He will even take on the Prince of princes in battle, but he will be broken, though not by human power."[323]

---

321 Mark 13: 21-23
322 Rev. 13:18
323 Daniel 8: 23-25

This biblical prophecy is recounted here not as an evangelical tool but merely to give context to the Lucis doctrine. The Lucis Trust has framed this debate, but has only given the public half the story. If they genuinely believe the story, they should explain the other half. Its followers claim the return of their light-bearing Coming One has been foretold by the prophets. You can make your own call as to how the story – which the UN aficionados claim is real, in their own words – is likely to end if you believe the prophecies.

"And I saw a great white throne, and I saw the one who was sitting on it. The earth and sky fled from his presence, but they found no place to hide."[324]

In assessing which version is most likely the correct interpretation of the text, consider this: The Lucis Trust, Theosophy, Buddhism, New Age all talk of hidden secrets revealed to initiates at different stages of their journey. They are essentially elitist, class-driven cults. Upper class aristocracy lording it over middle class bourgeois and lower class peasantry. These so called religions of 'tolerance' all argue the Bible is not meant to be understood by the general public, but only by "initiates" and "enlightened ones". Yet, the Jesus of the Bible came for the poor and the oppressed, so is it really likely that he would make his meanings so hidden to ordinary people that only a special elite could understand them?

There is something essentially dodgy, isn't there, about senior UN officials and their hangers-on praying to Lucifer to return and fulfil the prophecies? Especially when the prophecies make special mention of "the nations" that united against Christ:

"The armies of heaven, dressed in pure white linen, followed him on white horses. From his mouth came a sharp sword, and with it he struck down the nations."[325]

Pope Benedict when invited to the UN in 2008 was told by Ban Ki-moon:[326]

"The United Nations is a secular institution, composed of 192 States. We have six official languages but no official religion. We do not have a chapel – though we do have a meditation room."

Ban then invited Benedict to pray there, at the altar:

"Your Holiness, these are fundamental goals we share. We are grateful to have your prayers as we proceed on the path towards them. Before

324 Rev. 20:11
325 Rev. 19:14-15
326 "Remarks to introduce His Holiness Pope Benedict XVI to the General Assembly of the United Nations", Secretary-General Ban Ki-moon, General Assembly, 18 April 2008 http://www.un.org/apps/news/infocus/sgspeeches/statments_full.asp?statID=219#.UkkEcxCurV0

leaving the UN today, you will visit the Meditation Room. My great predecessor, Dag Hammarskjöld, who created that room, put it well. He said of the stone that forms its centerpiece [and I quote]: 'We may see it as an altar, empty not because there is no God, not because it is an altar to an unknown God, but because it is dedicated to the God whom man worships under many names and in many forms'."

Even the US *Catholic News* was fooled by it, reporting how the Meditation Room was an icon for the world.[327] The Pope, said the paper, wrote in the visitors' Golden Book at the Meditation Room. The symbolism of the head of the Catholic church appearing before the Lucis Trust's Lucifer-dedicated altar in the UN must have had the adepts chuckling. Even so, maybe Benedict suspected something. He had been scheduled to remain there for five minutes but, "emerged about a minute later walking briskly to the delegates entrance."[328] Maybe a minute inside Lucifer's temple was enough to tell the Pope where he was, enough to send a chill up his spine. Not only did he leave the Meditation Room rapidly, he left the entire UN building, cutting short his UN visit by a staggering 22 minutes.[329]

What did Pope Benedict know that you don't?

327 "U.N. room seen as reminder of world's need for prayer, meditation", Catholic News, 20 April 2008, http://www.catholicnews.com/data/stories/cns/0802195.htm
328 The Day, 19 April 2008, p3
329 Having said that, Pope Benedict had been a huge supporter of occultist priest Pierre Teilhard de Chardin, whose writings inspired the Luficerians and Robert Muller. Teilhard's writings and those of his supporter Henri de Lubac became part of the basis for the liberal Vatican II theology introduced by Popes John XXIII and Paul VI. In other words, Ratzinger was sympathetic to the theology of the suspect popes.

CHAPTER 21

The Day The Music Died

"And in the streets the children screamed
The lovers cried, and the poets dreamed
But not a word was spoken
The church bells all were broken
And the three men I admire most-
the Father, Son, and the Holy Ghost-
They caught the last train for the coast
The day the music died"

– Don McLean, American Pie

The UN Secretary-General might call his organisation "secular"; the facts show it is anything but. The United Nations sees itself on a "divine" mission to change the world, and nothing is going to get in its way.

Pope Benedict wasn't the first Pope to have contact with the occult team running the United Nations, however. We've already seen Pope John XXIII named as a co-founder of the New Age Temple of Understanding at the UN. Whether that was because John was spiritually naïve, or because it was deliberate, we may never know. What we do know is the occultist UN Secretary-General U Thant may have put the Pope up to it. It was Thant who also persuaded John's successor, Pope Paul VI, to become the first Pontiff to address the United Nations General Assembly in October 1965.

"One of Mr Thant's personal triumphs was the visit of Pope Paul VI to

UN Headquarters in 1965," reported newspapers.[330] "It was Mr Thant's idea...he often injected religious notes in his remarks. He maintained a warm friendship with both Pope John XXIII and Pope Paul, visiting them occasionally and exchanging messages with them.

"He believed that all ideologies should be able to coexist peacefully and that *they would eventually be synthesised.*" [emphasis added]

Pope Paul VI is also listed as a "founding friend" with U Thant and Pope John XXIII of the New Age Temple of Understanding,[331] although the extent of the two popes' involvement or understanding is not known.

In his lengthy address to the UN General Assembly, Pope Paul VI at times sounded like he was worshipping the United Nations itself:

"Gratitude to you, glory to you, who for 20 years have laboured for peace. Gratitude to you, glory to you for the conflicts which you have prevented...We feel that you thus interpret the highest sphere of human wisdom, and, we might add, its sacred character...Is there anyone who does not see the necessity of coming thus progressively to the establishment of a world authority, able to act efficaciously on the juridical and political levels?"[332]

Not only was Pope Paul the first Pontiff to speak at the United Nations, he was the first head of the Catholic Church to have prayed at an altar dedicated to Lucifer in the UN Meditation Room.

Under the subheading "Prays before block of iron", the *Boston Globe* records:[333]

"At the UN building on the East River front, Secretary-General U Thant, a Buddhist from Burma, met the sovereign of the church and walked at his side 50 yards through the vaulted lobby to the meditation room, where the Pope faced a plain block of iron ore, five feet high and two feet wide, bathed in a white light, with no inscription and no suggestion of any religious symbol, Christian, Jewish, Hindu, Moslem or any other.

"In the bare Meditation Room, the Pope and U Thant stood silently with closed eyes."

The *Globe* records the prayer at the iron ore altar as one of the "Holy Father's five great moments".

That was 1965. Fast forward nearly five decades and the New Age

330 "UN Secretary-General Leaves Office", *Toledo Blade*, 26 December 1971, p23
331 http://templeofunderstanding.org/who-we-are/founding-friends/
332 http://latimesblogs.latimes.com/thedailymirror/files/1965_1005_pope07.jpg
333 "A 14 Hour Triumph – a Day Without Parallel", by Laurence Winship, *Boston Globe*, 5 October 1965, p1

The Los Angeles Times   TUES., OCT. 5, 1965 **D**

World and U.N. Listen as Pope Paul Calls for Peace

A MESSAGE FOR THE WORLD—This was the scene in the U.N. General Assembly in New York as Pope Paul VI, at left in white robe, delivered his message—the high point of his day-long New York visit.

A Meeting on the Receiving Line

A WAVE TO NEW YORK—The Pontiff responds to greetings on arrival from Rome. Behind him is U.N. Secretary General U Thant, on hand for the welcoming.

AMONG THE GUESTS—Mrs. Jacqueline Kennedy is greeted by the Pontiff on receiving line following his address to U.N. General Assembly. In the center background is Sen. Edward Kennedy (D-Mass.).

THEY'RE ECSTATIC—The Pope gets an enthusiastic greeting as he arrives to visit the World's Fair.

SOVIET GREETING—Russian Foreign Minister Andrei Gromyko, right, meets Pope at the U.N. reception.

beliefs of the UN have permeated into every sector of society, and even into science.

"The world stands on the threshold of global change," said a group of scientists and "spiritual leaders" in an open letter to current UN Secretary-General Ban Ki-moon at the end of a New York conflab this year.[334]

"Ecological, political, anthropological, economic and other crises are intensifying. Wars are waged, resources wasted senselessly, and the planet is being polluted. Society is experiencing a crisis of goals and values. National leaders are concerned with short-term internal stability, yet pay insufficient attention to the problems and opportunities of the future of civilization. Human civilization essentially faces this choice: slide into

334 "22 leading scientists, technologists, entrepreneurs and spiritual leaders issue open letter to UN Secretary-General Ban Ki-moon", Globe Newswire, March 12, 2013, http://www.kurzweilai. net/globe-newswire-22-leading-scientists-technologists-entrepreneurs-and-spiritual-leaders-issue-open-letter-to-un-secretary-general-ban-ki-moon

Prays Before Block of Iron

How much time did he have to rest alone before he went to meet the President of the United States? Three hours sleep for the Pontiff the night before he left Rome. They say he does not sleep well on planes.

At the U.N. building on the East River front, Secretary-General U Thant, a Buddhist from Burma, met the sovereign of the church and walked at his side 50 yards through the vaulted lobby to the meditation room, where the Pope faced a plain block of iron ore, five feet high and two feet wide, bathed in a white light, with no inscription and no suggestion of any religious symbol, Christian, Jewish, Hindu, Moslem, or any other:

This plain stone had been the inspiration of Dag Hammarskjold, the earlier Secretary-General who weathered years of protests from world spokesmen that one religion or another was being favored.

In the bare Meditation Room, the Pope and U Thant stood silently with closed eyes.

The newest experience during any of his travels must have come to Paul VI when he was lifted 35 stories to the Goldberg apartment and its adjoining presidential suite at the top of the Waldorf Towers.

the abyss of global degradation, or realize a new model of development, *a model capable of changing human consciousness and giving new meaning to life.* [emphasis added].

"Global Future 2045 is premised on the notion that in order for civilization to move to a higher stage of development, *humanity vitally needs a scientific revolution and significant spiritual changes that will be inseparably linked,* supporting and supplementing each other."

Just how far has this New Age religious fundamentalism *spread* within official circles, and how did it happen?

In a 1995 briefing to mark the 50th anniversary of the founding of the United Nations, the Lucis Trust wrote that despite the bickering politics of the UN, "in its first 50 years the UN has been the pioneering force for unity in a divided world."

The Lucis Trust then notes, and this is important, that the UN's role in preparing for a one world government has been essential, and successful:

"Through legal agreements, innumerable meetings and a vast array of action-oriented programmes, *the UN system has established an architecture and a political culture within which a new interdependent world order can now take shape.*" [emphasis added]

The UN's propaganda against the public using NGOs and the news media has been "enormously influential in shaping global opinion and ultimately, action," says the Lucis Trust.

"It is the awakening amongst 'we the peoples' of a will for action based on a recognition of human and planetary unity that can be expected to characterise the next chapter in the story of the UN. People's movements are beginning to play an unprecedented role in national and global affairs. They are now key players in the political scene and there is ample evidence to suggest that the powerful environmental and development bodies are ready to mobilise support for government action in line with UN objectives."

As a further illustration of just how much impact the Satanists have had on UN organisations, a senior World Bank official named Richard Barrett addressed the Lucis Trust's World Goodwill conference in 1996.

Barrett revealed he had been invited to establish a "spiritual unfoldment[335] society" at the World Bank, based on New Age and Alice Bailey doctrine. Very quickly, the meetings were attracting 40 to 50 World Bank HQ staff at a time, and "within a few months" the Luciferian offshoot was so "perfectly respectable" within the Bank that "we began to announce our meetings on the internal email system."[336]

Barrett then hatched the idea of the World Bank hosting "an international conference on spiritual values and sustainable development".

That conference, staged at the World Bank on 2-3 October 1995, attracted more than 400 delegates from around the world. As the professional journal *Environmental Conservation* reported in its March 1996 issue, the gathering attracted "diplomats…eminent environment and political scholars, ethicists and theologians…NGOs" and a range of other hangers-on. All of them being subtly indoctrinated with Lucis Trust teachings.[337]

No less a personage than the head of the World Bank, James Wolfen-

335 An occult term meaning progressive revelation, coined by Satanist Helena Blavatsky in "Deity, Cosmos & Man", part 2, Chapter 11, http://www.blavatskytrust.org.uk/html/deity%20cosmos%20 and%20man/dcm%20p2%20ch11.htm
336 http://web.archive.org/web/19990224010437/http://www.lucistrust.org/goodwill/wgnl972.htm
337 "Ethics and spiritual values etc." Arthur Westing, Environmental Conservation / Volume 23 / Issue 01 / March 1996, pp 89-89 DOI: http://dx.doi.org/10.1017/S0376892900038297

sohn, acknowledged that spiritual values would henceforth guide World Bank policy and decisions.[338]

Barrett revealed that since the international conference, the World Bank appointed him to "the position of Values Coordinator in the newly formed Department of Institutional Change. We have created a Values Circle... the greatest need we have, right now, is for each of us who cares about the future of humanity and the planet to bring this consciousness into our workplaces and begin the process of transforming our organisations."[339]

In short, the infection has jumped from the United Nations to the World Bank as well.

In truth, it has gone a lot further than that, as you're about to see.

---

338 Ibid
339 World Goodwill Newsletter 1996, http://web.archive.org/web/19990224010437/http://www.lucistrust.org/goodwill/wgnl972.htm

CHAPTER 22

Control Of The Churches

"Have you ever stood and stared at it, marveled at
its beauty, its genius? Billions of people just living out
their lives, oblivious. Did you know that the first Matrix
was designed to be a perfect human world, where
none suffered, where everyone would be happy? It
was a disaster. No one would accept the program."
– Agent Smith, The Matrix

Nearly every major corporate and government agency gets visited these
days by New Age "consultants" using the cloak of corporate team-building
and consultancy to preach the same consciousness-raising, common-goal
doctrine. Remember, as the elite of the New Age, Bailey and Blavatsky,
made clear, it is the raising of "consciousness" worldwide that they think
will usher in Satan's return. They don't particularly care how consciousness is raised, as long as it is. There is huge money being made in the
New Age movement through these speaking tours.

You may well have attended some of these seminars in your own work-
place. Your department's code of practice might now include many of
these themes.

What might shock you is how many of the Christian churches, Catholic,
liberal Protestant and even Pentecostal Protestant, are deeply infested with
the Lucifer doctrine. Again, regardless of what you personally believe,
the key players in this do understand it, and to help you understand the

narrative here, I need to give you a little more background.

In 1990, Vatican priest, intelligence analyst and professor, Father Malachi Martin, published an epic 735 page non-fiction bestseller called *The Keys Of This Blood*, covering some of the issues raised here. Back in 1990 much less was known; for example, Martin knew nothing of the Lucifer altar inside the UN, and because he was unaware of its significance he evidently missed the real significance of Pope Paul VI praying at the Satanic altar. However, what he did know were some of the events taking place inside the Vatican itself:

"Most frighteningly for [Pope] John Paul [II], he had come up against the irremovable presence of a malign strength in his own Vatican and in certain bishops' chanceries. It was what knowledgeable Churchmen called the 'superforce.' Rumors, always difficult to verify, tied its installation to the beginning of Pope Paul VI's reign in 1963. Indeed Paul had alluded somberly to 'the smoke of Satan which has entered the Sanctuary' – an oblique reference to an enthronement ceremony by Satanists in the Vatican. Besides, the incidence of Satanic pedophilia—rites and practices— was already documented among certain bishops and priests as widely dispersed as Turin, in Italy, and South Carolina, in the United States. The cultic acts of Satanic pedophilia are considered by professionals to be the culmination of the Fallen Archangel's rites."[340]

The book was published well before the issues of priests and pedophilia really hit the headlines, but it is interesting that Martin ties the rise of Satanism and pedophilia within the Vatican to the early 1960s and the transition between John XXIII and Paul VI.

The Keys Of This Blood was a non-fiction book. Martin later told interviewers he had to be careful how much he revealed in that. He was much harder hitting in a follow-up book written, on the face of it, as a fictional novel. In *Windswept House*, Father Martin returned to the scene of the crime, and this time directly alleged – albeit disguised as 'fiction' – that Satan had formally been "enthroned" in the Vatican:

"The Enthronement of the Fallen Archangel Lucifer was effected within the Roman Catholic Citadel on June 29, 1963; a fitting date for the promise about to be fulfilled. As the principal agents of the Ceremonial well knew, Satanist tradition had long predicted that the Time of the Prince would be ushered in at the moment when a Pope would take the name of the

---

340 The Keys Of This Blood, by Malachi Martin, Simon & Schuster NY, 1990, p632

Apostle Paul. That requirement – the signal that the Availing Time had begun – had been accomplished just eight days before with the election of the latest [Pope Paul VI]."[341]

Father Martin wrote that the enthronement ceremony had to be the opposite of what Jesus Christ had gone through and taught.

"If the aim was to be achieved – if the Ascent of the Prince was actually to be accomplished in the Availing Time – then every element of the Celebration of the Calvary Sacrifice [Jesus' death on the cross] must be turned on its head by the other and opposite Celebration.

"The sacred must be profaned. The profane must be adored. The unbloody representation of the Sacrifice of the Nameless Weakling on the Cross must be replaced by the supreme and bloody violation of the dignity of the Nameless One. Guilt must be accepted as innocence. Pain must give joy. Grace, repentance and pardon must all be drowned in an orgy of opposites. And it must all be done without mistakes. The sequence of events, the meaning of the words, the significance of the actions must all comprise the perfect enactment of sacrilege, the ultimate ritual of treachery."

Because the Vatican, even in 1963, had high security, the planners realised that certain trappings of the Satanic rite could not escape detection if they were carried out within the Vatican chapel itself, so they arranged for a mirror ceremony to be held in the United States at precisely the same time. That ceremony, conducted by telephone link simultaneously, would include the things that priests in the Vatican could never be caught with, including a young girl – the daughter of one of the men involved – who was drugged and raped on the altar being used in the US. Her hands and feet were also pricked to generate blood in the places Christ was pierced, and the blood used in the ritual, reported Father Martin.

The child had to be sexually violated, as a representation of "The Virgin raped" for the purposes of the ritual.

The general form of the prayers, you may recognise:

"I believe in One Power, and its name is Cosmos. I believe in the Only Begotten Son of the Cosmic Dawn, and His Name is Lucifer. I believe in the Mysterious One."

Did this event, originally discussed in *Keys Of This Blood* and then explicitly detailed in *Windswept House*, actually happen?

---

341 Windswept House by Malachi Martin, Doubleday NY, 1996, p7

"Oh yes, it is true; very much so," Martin told *New American* magazine. "But the only way I could put that down into print is in novelistic form."

Another Catholic Archbishop caused a storm of controversy in the Italian press in 1996 when he too alleged Satanism was being actively practiced by priests and even higher ranks within the Vatican's walls. Attached to the Holy See to work on immigrant issues, Archbishop Milingo said: "Certainly there are priests and bishops. I stop at this ecclesiastical level of hierarchy. I am an Archbishop, so higher than this I cannot go."[342]

When challenged by journalists, he reminded them of the late Paul VI's comment about the smoke of Satan entering the Vatican, and added, "No one since that time has ever said that he has exited from there."[343]

Malachi Martin, for his part began working on a new book in 1998 entitled *Primacy: How The Institutional Roman Catholic Church Became A Creature Of The New World Order.* He died in unusual circumstances before he could complete the book. It has never been published.

"A tragic fall, reportedly delivered by 'an unseen hand', caused Malachi Martin's second stroke in twelve months," reported Unity Publishing in the US.

"The invisible (preternatural?) force that shoved Father Martin into a stumble, wherein he hit and fatally traumatized his head remains unknown. Yet, before an accompanying stroke claimed his physical existence, while lying in critical condition, Father managed to convey to a close friend, prudently preferring to remain anonymous: 'I felt something push me, but... no one was there'."

Martin had also accused the two popes we met in the earlier chapter, Paul VI and John XXIII, of being involved in freemasonry, a movement whose higher theology is closely aligned to the Lucifer doctrine.[344]

Readers may remember British author David Yallop's 1980s bestseller *In God's Name*, in which he alleged Pope John Paul 1 was murdered, and that the Vatican Bank was run by Catholic freemasons belonging to the P2 Lodge. When key figures in the bank turned up hanging from bridges or sprayed with bullets in drive-by executions, that only heightened specula-

---

342 *Il Tempo*, 27 November 1996
343 *La Nazione*, 27 November 1996
344 Les Amis du Christ-Roi (1997), L'Eglise Eclipsée? Réalisation du complot maçonnique contre l'Eglise. Témoignage inédit du père Malachi Martin, présent en qualité d'intreprète aux derniers Conclaves [The Church eclipsed? Realisation of the masonic conspiracy against the Church. Inedit testimony of Father Malachi Martin, present as an interpreter at the last Conclaves] (in French), Dinard: Delacroix, ISBN 2-9511087-0-2

tion about what kind of people were seemingly in control of the Vatican.

That issue has come back to haunt the latest incumbent, Pope Francis, who remarked soon after taking power about the emerging influence of freemasons:

"A secret society of Masons, so many secret societies. This is the most serious problem for me," he told the press this year.[345]

One who knew about freemasonry's stranglehold on the Vatican only too well is Father Luigi Villa, a doctorate of theology who was tasked by Pope Pius XII, no less, with investigating Masonic influence within the Catholic Church. Born February 1918, Villa died November 2012, just five years short of his 100th birthday. He left behind a book documenting what he found. Trawling through more than 30,000 papal documents, Villa discovered a Pope in Paul VI who spoke in the coded masterful and light-filled language of the Lucis Trust:

"We fail to perceive, instead, that we are the masters of life already. Doubt has entered our conscience, and it has entered through windows that were supposed to be opened to the light instead ..."Paul VI said.[346]

In a speech a decade earlier, heralding the introduction of the liberalising Vatican II reforms, Paul VI also talked of "Jesus the Master", and then added, "if the whole world is changing, will not religion change as well?"[347]

A hint of how far gone Pope Paul was can be seen in the first of four fundamental conditions he laid out for the future work of the Church:

"Clarity: which should consist in a perfect balance of position between the two dialoguing parties."[348]

As Villa pointed out in his warnings to the Vatican, that statement suggests the Church should find a compromise with those challenging it, rather than maintain a stance that might be seen as "exclusive". The Truth, said Villa, should never be agreed by compromise. It either is the truth, or it isn't. Christ, said Villa, had ordered his disciples to "preach" to all nations, not compromise with all nations and knock out the bits of the gospel the other side didn't agree with.

For two thousand years, he said, the Church had maintained itself a

---

345 "Most of us would laugh at the idea of a masonic mafia at work in the Vatican. I'm not sure that we should", by Father Alexander Lucie-Smith, Catholic Herald UK, 30 July 2013, http://www. catholicherald.co.uk/commentandblogs/2013/07/30/most-of-us-would-laugh-at-the-idea-of-a-masonic-mafia-at-work-in-the-vatican-im-not-sure-that-we-should/
346 Speech, 29 June 1972, quoted in Paul VI Beatified? By Fr Luigi Villa
347 Paul VI Beatified?, p23, see entire book at http://www.huttongibson.com/PDFs/Paul-VI-Beatified-Book.pdf
348 "Ecclesiam Suam" from Vatican II, cited in ibid, p26

rock in human affairs, but now Popes John and Paul in setting up Vatican II had cut the Church adrift and ordered compromise. The Church, argued Paul obliquely, no longer existed to serve God: "All this doctrinal wealth points but to one direction: to serve man. The Church has, so to say, declared Herself the servant of Humanity.

"To know God, one has to know man," said Paul, in almost a direct rip-off of the Lucis Trust theology. Even people who haven't darkened a church doorway in decades might recall that the central tenet of Christianity was humankind's fallen state leaving us anything but godlike. If 'we' are God, how do we explain murder, rape, cheating and war?

Pope Paul VI talked of a brave new world to come and a brotherhood of all men from every nation:

"All of us, Churches included, are involved in the birth of a new world. God... in His love for man, organizes the movements of history for the progress of humanity and in view of a new earth and new heavens, wherein justice shall be perfect.

"The Catholic Church urges all of her sons to undertake, together with all men of good will of every race and nation, this peaceful crusade for the well-being of man... in order to establish a global community, united and brotherly."[349]

It is a subtle statement that you might miss if not pointed out more directly: the central thesis of Christianity is that Jesus Christ returns to earth to establish his Christian world – at least, that's how the Bible tells it. But in those passages above Paul speaks in the present tense of a human-driven peace "crusade" not led by but "including" the Churches. Again, regardless of whether you believe in Christianity or not, you can appreciate that it is a fundamental departure from the Bible version of events. It is also exactly what the Lucis Trust were preaching.

"Isolation is no longer an option," said the Pope. "The hour has come of the great solidarity among men, toward the establishment of a global and fraternal community."

This was the Pope, of course, who prayed at the altar of Lucifer in the United Nations building. The Lucis Trust talked, as we saw earlier, of the need to raise consciousness in order to boost "the Ascent of Man" and call in 'the Coming One'. Here's how Pope Paul VI says pretty similar things:

"The Church, although respecting the jurisdiction of the Nations, must

---

349 Ibid p63

offer her help to promote a global humanism, I mean to say, an integral development of man as a whole and of each and every man... Placing herself at the forefront of social action, She must direct all of her efforts to sustaining, encouraging, and driving the initiatives that operate toward the integral promotion of man."[350]

By now, Pope Paul VI was singing directly from the Satanistic handbook, making speeches from the Vatican that almost precisely mirrored what the occultists were endeavouring to do:

"Before being a policy, *peace is a spirit... It forms*, it takes hold of the consciences, in this philosophy of life each has to build for himself, *as a light* for his steps upon the paths of the world and in the experiences of life. That means, dearest brothers and sons, that peace requires an education."[351]

Everyone on earth, believer, non-believer, apparently requires the same thing: to be educated about peace by the light-bringers.

If you are a Roman Catholic and hoping that this has all been misinterpreted, that the Pope meant the Peace of Christ, a god-given peace, think again:

"A peace that is not born of the real cult of Man, is not essentially a peace," said the Pope.[352]

Only Man's peace is real? The Pope's words directly clash with the words of Jesus Christ, who's quoted as saying, "Peace I leave with you; my peace I give you. I do not give to you as the world gives."[353]

Regardless of whether you believe in Christianity or not, it should be clear that the head of the Catholic Church was saying something that contradicted his boss. That, alone, should be a dead giveaway.

After praying at Lucifer's iron-ore altar at the UN on October 4, 1965, Pope Paul VI then swore allegiance to the United Nations on behalf of the entire Catholic Church and all people on earth:

"Our message is meant to be, first of all, a moral and solemn ratification of this lofty Institution... We bring to this Organization the suffrage of Our recent Predecessors, that of the entire Catholic Episcopate, and Our own, convinced as We are that this Organization represents the obligatory path of modern civilization and of world peace... The peoples

350 Ibid, p64
351 Ibid, p69
352 Ibid, p69
353 John 14:27

of the earth turn to the United Nations as the last hope of concord and peace. We presume to present here, together with Our own, their tribute to honor and of hope."[354]

Adding impact to this show of homage to the United Nations, the head of the Catholic Church then gave UN Secretary-General and Lucis Trust follower U Thant the symbols of his papal authority, "his pastoral and his ring", writes Father Villa to the Vatican.

"On November 13, 1964, he would [also] remove the tiara (the triregno) on the altar, definitively renouncing it," he added.

What could be more symbolic of a religious tipping point: the Pope himself praying at a pagan altar, not just any pagan altar but one officially run by self-proclaimed disciples of Satan, a Pope who then swears the allegiance of his entire Church to the United Nations (the same UN with Lucifer's temple at its heart), and who then hands the head of that UN his symbols of papal authority.

Some years before he became pope, Paul VI discussed the issue of freemasonry with a fellow priest, who later wrote about it. According to Father Felix Morlion, the future pope told him:

"Not a generation will pass and, between the two societies (Church and Freemasonry), peace shall be sealed."[355]

The deal was indeed sealed. In 1967 the Catholic Church allowed its Scandinavian Catholics who had converted from protestantism to be active freemasons also, and the move was further extended under Papal seal in 1974.

One senior French mason wrote in his book of the time, "One could really speak of a Revolution that from our Masonic Lodges has spread out magnificently, reaching the top of St. Peter's Basilica."[356]

Little wonder that Italy's freemasons paid tribute to Pope Paul VI when he died:[357]

"It is the first time – in the history of modern Freemasonry – that the Head of the greatest Western religion dies not in a state of hostility with the Freemasons! ... For the first time in history, Freemasons can pay respect at a Pope's tomb, without ambiguity or contradiction."

---

354 Address of Pope Paul VI to UN General Assembly, 4 October 1965
355 Paul VI Beatified?, p98
356 "Ecumenism From the Perspective of a Freemason of Tradition," Msr Marsaudon, cited in Ibid at p100.
357 "Obituary for Paul VI, by Grand Master of Palazzo Giustiniani [Rome headquarters, Grand Orient of Italy], Giordano Gamberini, *La Rivista Massonica* magazine, 1978

Much has been written by others about the rise of the occult within the Catholic Church, and this chapter is not intended to be exhaustive on the point.

Suffice to say that the spiritual drive toward totalitarianism that emanated from within the United Nations has clearly been shown to have implanted itself in the very heart of the world's largest Church as well.

As if to prove the point, just as this book was entering its final edit, Pope Francis hit the headlines:

"Pope Francis cranked up his charm offensive on the world outside the Vatican on Tuesday, saying in the second widely shared media interview in two weeks that each person 'must choose to follow the good and fight evil as he conceives them' and calling efforts to convert people to Christianity 'solemn nonsense'.[358]

"The Vatican's head seemed intent on distancing himself from its power, saying church leaders 'have often been narcissists' and 'clericalism should not have anything to do with Christianity'.

The idea that the public are the final judge on what is good or bad is an abandonment of the Church's position on all moral issues. Pope Francis appeared to be clear he meant it:

"Asked if there is a single vision of good, and who decides, Francis says:

" 'Each of us has a vision of good and of evil. We have to encourage people to move towards what they think is good ... Everyone has his own idea of good and evil and must choose to follow the good and fight evil as he conceives them. That would be enough to make the world a better place'.

"Asked if he feels touched by grace, Francis tells the atheist reporter that the holy quality 'is the amount of light in our souls, not knowledge nor reason. Even you, without knowing it, could be touched by grace'."

And there's the money quote: the holy quality "is the amount of light in our souls" – something the light-bearer worshippers will be delighted to hear, but a considerable distance from the biblical version.

The suggestion that converting people to Christianity is "solemn nonsense" is one that many non-religious readers might agree with, but it is not what Jesus Christ preached:[359]

"It was written long ago that the Messiah must suffer and die and rise

358 "Pope Francis stirs debate yet again with interview with an atheist Italian journalist", Washington Post, 2 October 2013
359 Luke 24:46-47

again from the dead on the third day. With my authority, take this message of repentance to all the nations, beginning in Jerusalem: 'There is forgiveness of sins for all who turn to me'."

Not enough for you, need more proof?

"Jesus came and told his disciples, 'I have been given complete authority in heaven and on earth. Therefore, go and make disciples of all the nations, baptizing them in the name of the Father and the Son and the Holy Spirit. Teach these new disciples to obey all the commands I have given you. And be sure of this: I am with you always, even to the end of the age'."[360]

What was it that Pope Francis actually said?

"Proselytism [evangelising and baptising] is solemn nonsense, it makes no sense."[361]

On the strength of that, it is probably fair to say the Catholic Church

---

360 Matt 28:18-20
361 http://www.christianpost.com/news/pope-francis-in-interview-with-famous-atheist-identifies-worlds-most-serious-evils-calls-proselytism-nonsense-105754/

is continuing down the path begun by Popes John XXIII and Paul VI. If Satan entered the Sanctuary in 1963, all the evidence now suggests the popes have become little more than sock-puppets. If you are not religious, and the symbolism is lost on you, simply take it as read that if a Church's doctrine goes against what Jesus Christ stated in the Bible, then on the biblical definition it's what's called a "false gospel", and a false church.[362]

Any pope abandoning the instruction to evangelize has gone AWOL at his command post, either fundamentally misunderstanding his orders, or as a result of treason.

Little wonder, say some, that a lightning bolt struck St Peter's Basilica at the Vatican the night Pope Benedict resigned,[363] the first pope to do so in 600 years and an event that paved the way for the ascension of Francis.

This is not an attack on Catholicism or Catholics. It would be unfair, therefore, to leave this without also mentioning the protestant churches. Most of them, too, have long since surrendered. The Anglican/Episcopalian Church is barely clinging to the Bible by a toenail, and the other mainstream denominations are not far behind. Some Archbishops of Canterbury have been senior freemasons, so the conflict of interest manifested a long time ago.

In fact, many of the world's main denominations have taken up with the United Religions Initiative (URI), established by an Anglican bishop in California at the insistence of the UN's Robert Muller. The essence of the URI is "interfaith" negotiations with other religions, to come to some sort of compromise faith statement the world can agree on.

Similar moves are afoot with the UN's Alliance of Civilisations project, which aims to create a global interfaith compromise.

Deserving particular mention, however, are some of the huge Pentecostal churches that have sprung up, like Rick Warren's purpose-driven Saddleback Church in Lake Forest, California, or Bill Johnson's Bethel Church in California behind the "Manifest Presence" movement, or Joel Osteen Ministries in Texas. There are many others, but these examples are typical cases.[364]

---

362 "Your minds may somehow be led astray from your sincere and pure devotion to Christ. ⁴For if someone comes to you and preaches a Jesus other than the Jesus we preached, or if you receive a different spirit from the Spirit you received, or a different gospel from the one you accepted, you put up with it easily enough," warned the apostle Paul in 2 Cor. 11:3-4
363 "Lightning bolt strikes the Vatican's St Peter's Basilica – video", The Guardian, 12 February 2013 http://www.theguardian.com/world/video/2013/feb/12/lightning-vatican-peter-basilica-video
364 All three of these churches are guilty of falsifying Christian doctrine by publishing statements and claims not supported in the Bible and in many cases directly contradicting it. That is a debate

The thing common to many supposedly Christian ministries is that when you scratch the surface, it's simply the New Age in Christian drag. In fact, when you scratch a little deeper, it's the doctrine of the Lucis Trust staring right back at you.

Case in point?

Pastor Rick Joyner wrote praising an old time evangelist named John G. Lake. Lake is extremely well known in Pentecostal circles, and was preaching around the same time that Alice Bailey was establishing the Lucifer Publishing Company and the Lucis Trust.

Here's what Joyner said of Lake:

"Recently," says Joyner, "I was given a little book of sermons by John G. Lake, one of the great Christian leaders of modern times. Lake not only had one of the most effective ministries in history, he seemed to be one of those rare souls who walked in a manner worthy of his calling to the end, never falling to the pride or other worldly seductions that trip up so many. He wrote shortly before he died in 1935:

" 'I am getting ready in my soul, not to preach the old message with the old fire, but the new message with the new fire – not only to ask men to be good and go to heaven when they die, but to be God-like in character, nature, substance, and being. God is priming our souls. He is going to send forth the living waters and call by experience the new

---

for another time, however. The reference here is to their links with New Age symbols and doctrine.

order of Knights of the Morning Star into being and action'." (*Spiritual Hunger and Other Sermons* by John G. Lake, page 18. Published by Christ For The Nations, Inc.)[365]

Two things emerge from Lake's words, as approvingly cited by Rick Joyner. Firstly, that he has *not* come to preach the "old message" – salvation through Christ. Instead, he has "the new message" – to be "god-like in…nature, substance and being". We've heard that before.

Remember that quote from Genesis where Satan tempted Eve with the promise that if she ate from the Tree of Knowledge, "ye shall be as Gods"? It was the very first temptation offered to humans in the Bible, and here it is being offered by both the Lucis Trust and a supposedly Christian evangelist.

Secondly, you will have noticed the "Knights of the Morning Star". The passage from the Book of Isaiah in the Bible says the Morning Star was Lucifer. This deliberate fusion of New Age and Satanism into Pentecostal Christianity is no accident. No genuinely discerning Christian preacher should ever use the ambiguous phrase Morning Star, and if you are in a church that does, run for the hills. You may recall some of the passages from the Lucis Trust and others, emphasising that words have power and must be chosen very carefully. The "Knights" of anything tended to be Masonic in origin, so that's another giveaway.

Why would a supposedly Christian evangelist – two of them in fact, if you include Rick Joyner's endorsement[366] – call into being a new order of the Knights of the Morning Star, when the name meant 'Lucifer'? Surely only a preacher who thought his flock were idiots would pull such a stunt.[367]

---

365 http://web.archive.org/web/20040626110335/http://www.eaglestar.org/pages/questions/answer_index/Oct_1_2001_knights.html

366 Rick Joyner's operation is also called "Morning Star Ministries". Joyner would argue he means Jesus Christ, but his theology suggests he doesn't. He preaches exactly the same Coming One's Kingdom as the Lucis Trust, daring to go so far as to say it will change the definition of Christianity: "What is about to come upon the earth is not just a revival, or another awakening; it is a veritable revolution. The vision was given in order to begin to awaken those who are destined to radically change the course, and even the very definition of Christianity. The dismantling of organizations and disbanding of some works will be a positive and exhilarating experience for the Lord's faithful servants. A great company of prophets, teachers, pastors and apostles will be raised up." – Rick Joyner, *The Harvest*. Again, Joyner like Pope Francis contradicts his supposed boss, who said that if anyone changed the meanings of the Book of Revelation, "God will remove that person's share in the tree of life". You will see just how wrong the Kingdom Now/Revival claims are later in this chapter.

367 Whilst it is true that the Bible in places also refers to Jesus as the morning star, the context of John G. Lake's occult symbolism and his theological error leaves little room for doubt as to which Morning Star he meant.

Lake, incidentally, preached his gospel twice every Sunday, not from a church but from the Masonic Temple in Spokane, Washington.[368] Spokane is now home to the International Association of Healing Rooms, which is based in the same occult traditions.

Lake's website uses occult imagery, such as the Knights Templar symbol of the sword through the crown, and also a variation on the occult Pentagram. His suggestion that believers should become Knights of the Morning Star is proof that, just like the Catholic Church, modern protestant Christian evangelism has become heavily tainted.

A freemasonry website run by a 32nd degree Master Mason explains the significance of the pentagram to Masons:[369]

"This symbol then is a representation of humanity, and our Divine role in the Universe as co-creators of eternity. In addition to being a central altar piece in our Rosicrucian Temples, and the Blazing Star of the Craft Lodges, the Pentagram appears as an ensign in some of the High Degrees and rites. For example, it is central on the apron of the 28th Degree of the Ancient & Accepted Scottish Rite, *Knight of the Sun or Prince Adept.* In discussing the symbol of the pentagram in the lecture of this degree, Pike writes in *Morals & Dogma* that: 'in certain undertakings [the Pentagram] cannot be dispensed with. It is what is termed the Kabalistic pentacle... This carries with it the power of commanding the spirits of the elements.'

"A central lesson of this highly Kabalistic and Alchemical degree is that there is no death, only change. The Pentagram, symbol of humanity as the microcosm is an apt representation of this wisdom...It may be said that the Pentagram represents the power of the Divine Will, as manifested in Humanity, to effect conscious change. As conscious participants with the Divine Will, humanity is in the unique position of being able to be co-creators with the Divine."

Why would genuine Christian evangelists be donkey-deep in Masonic and occult imagery and doctrine? You probably know the answer by now.

---

368 http://www.enterhisrest.org/charismata/lake_biography.pdf
369 http://www.freemasons-freemasonry.com/pentagram_freemasonry.html

THE MAGICIAN.

It gets even better, for those who recognise occult symbols. The pentagram includes a geometric proportion long said to have special symbolism:

"The occult interpretation of this mathematical proportion is sometimes expressed in the maxim: 'Nature is to man as man is to God.' The proportion itself is often called the Golden Section of Pythagoras, and is associated with the Hermetic axiom: 'That which is above is as that which is below, and that which is below is as that which is above'."

Hermetic thought is a branch of occultism dating back to the ancient Greeks, and what you have just read in that last line is one of its fundamental spells – As above, so below. As below, so above:

" 'As above, so below' – a 'great word', a sacramental phrase, a saying of wisdom, an aphorism, a mystic formula, a fundamental law," wrote occultist and Theosopher G R S Mead in 1919.[370]

The concept can be seen throughout the New Age movement, but is most beautifully illustrated in the Ryder-Waite Tarot of the 19th century, where The Magician demonstrates the spell, "as above, so below":

The last place you would therefore expect "as above, so below" to show up is in the Bible. It doesn't, of course, well, not in the legitimate ones, but it does in the Bible version by Eugene Peterson known as *The Message*, used extensively by Rick Warren and Bill Johnson, sold in 'Christian' bookstores and used by millions of Christians worldwide.

This is the Lord's Prayer, as used in The Message (Matt 6:7-13): [*original King James version in brackets for comparison*]

Our Father in heaven, [*Our Father, which art in heaven*]

Reveal who you are. [*Hallowed be thy name*]

Set the world right; [*Thy kingdom come, thy will be done*]

Do what's best— as above, so below. [*In earth, as it is in heaven*]

Keep us alive with three square meals. [*Give us this day our daily bread*]

Keep us forgiven with you and forgiving others. [*And forgive us our debts, as we forgive our debtors*]

370 http://www.gnosis.org/library/grs-mead/grsm_asabove.htm

Keep us safe from ourselves and the Devil. [*Lead us not into temptation, but deliver us from evil*]

You're in charge! [*For thine is the kingdom*]

You can do anything you want! [*And the power*]

You're ablaze in beauty! [*And the glory*]

Yes. Yes. Yes. [*Forever and ever, Amen*]

Suddenly, what had been "Thy will be done in earth as it is in heaven" has been changed to an occult phrase that means all of creation is God, and vice versa. God is no longer a person but a 'thing' for humans to control. How did *that* happen?

The whole point of the New Testament in real Bibles is that "our Father in heaven" is revealed as Jesus Christ. Yet *The Message* is asking this mystery figure "Reveal who you are". Which "Father" is *The Message* really urging you to pray to? A clue is found in "you're ablaze in beauty, yes, yes, yes!". Apart from sounding like a rip off of the "I'll have what she's having" TV commercial, where in the Bible do we find fiery beauty?

The UN's Robert Muller told people to read the prophets, but if you do you will find it is written: "I drove you in disgrace from the mount of God, and I expelled you, O guardian cherub, from among the fiery stones. Your heart became proud on account of your beauty, and you corrupted your wisdom because of your splendour. So I threw you to the earth." (Ezk.28:16-17)

Odd, don't you think, that the Lord's Prayer, using the occult "as above, so below" incantation, is now offering praise to the beauty of someone as yet *not* revealed? Who might that be, and why do Rick Warren and Bill Johnson preach from *The Message*?

It could be for their new book, *The Seven Habits of Highly Effective Heretics*, but it's more likely something else.

When Johnson and Warren pray, "Our Father in heaven, reveal who you are", what spirit are they really asking to come to Earth?

It is written:

"Anyone who has seen me, has seen the Father," said Jesus. (John 14:9)

How could a genuine Christian pastor who was familiar with the Bible offer up a prayer to God asking *"Reveal* who you are"? Even atheists like Richard Dawkins reckon the Bible is pretty up front about who God is. So, again, who are Bill Johnson and Rick Warren really asking to reveal himself?

In 2 Thessalonians, we are told a rebellion in the faith will happen close to the end of times, at which point "the man of lawlessness is *revealed,* the man doomed to destruction. He will oppose and will exalt himself over everything that is called God or is worshipped, so that he sets himself up in God's temple, proclaiming himself to be God. (2 Thess. 2:3-4)

"The coming of the lawless one will be in accordance with the work of Satan, displayed in all kinds of counterfeit miracles, signs and wonders, and in every sort of evil that deceives those who are perishing. They perish because they refused to love the truth and so be saved. For this reason God sends them a powerful delusion so that they will believe the lie and so that all will be condemned who have not believed the truth but have delighted in wickedness." (2 Thess. 2:9-12)

Additionally, it is written: "Take heed that no man deceive you. For many shall come in my name, saying 'I am Christ'; and shall deceive many." (Matt 24:4-5)

Author Warren Smith, in his seminal bestselling book – *Deceived on Purpose* – on what he called the evil of Rick Warren's Saddleback Church, describes how he woke up to the deception. Smith describes how esoteric writers try and subtly steer you off course, and uses as an example *The Message* 'bible's alternative version of the words from Jesus you just read:

"Jesus said, 'Watch out for doomsday deceivers. Many leaders are going to show up with forged identities, claiming 'I am Christ, the Messiah.' They will deceive a lot of people."

As Smith points out, the removal of the word "man" from the original words of Christ, and replacement with the "doomsday deceivers", sets followers on a totally different look out. Instead of looking warily at anyone who compares themselves to Jesus, *Message* readers are looking for "leaders" promoting "doomsday". The prosperity gospel movement doesn't preach doomsday, it preaches love, peace and abundance for all in the oneness of the Christ. In fact, *Message* readers would be more likely to see this book you are now reading, or Smith's, as examples of "doomsday deceivers".

Talk about bait and switch. Smith says it hides the consequences of deception from those foolish enough to believe it. "My wife and I were not 'doomsday deceivers'. We were not 'leaders with forged identities'. If we had been looking only at Eugene Peterson's paraphrase when we were unbelievers, we would have never seen ourselves and the whole New Age movement in that prophetic passage of Scripture. But thanks to a *real* Bible, we were clearly shown that we were the subjects of Jesus' warning."

But hang on, some of you are saying. That *Message* version of the Lord's Prayer did ask to keep us safe from the Devil. Well yes, until you read New Age dictionaries that define 'Devil' as "the mass of thoughts...that fight against the truth",[371] or define it as "the flesh and its desires".[372] The Devil is no longer an entity, but merely a state of mind. In this way, the New Age manages to cloak its doctrine in Christian terminology. The "Coming One" is referred to as "the Christ", evoking a knee-jerk lizard-brain subconscious assumption in people, and the "Devil" is referred to as well, provoking a similar subconscious assumption. Only the adepts, those who know how to read New Age code, really understand the significance.

Rick Warren quotes extensively from *The Message* as his preferred 'bible' in *Purpose Driven Life*. You will find the "above and below" incantation on the first page of his first chapter.

Rick Warren also speaks approvingly of New Age religious leaders. Bernie Siegel is praised as a man with a "deep sense of life purpose" in chapter three of *The Purpose Driven Life*. Warren doesn't disclose that Siegel is a New Age guru.

Warren Smith does a sterling job of exposing the New Age-sourced false preaching of Rick Warren and his heretical mentor at the Crystal Cathedral, Robert Schuler, so I won't repeat any more of it here, instead urging you to read his book *Deceived on Purpose*.[373] There are simply too many false gospel pastors and evangelists loose in the world, and writing a book on them all would be like playing 'whack a mole'.

You will recall Bethel Church's Bill Johnson and the "Manifest Presence" movement he leads. Now read this extract from a century-old occult Freemasonry text:

"Their disciples are the forward-guard, the forerunners of the *manifest presence of the Master* later to come."

Say what? 'Manifest presence' is a Freemasonry term as well? What's the context of that passage? It turns out to be a document revealing how the "Master" – a coy reference to the One, or "Lucifer" – is working through a number of religious entities to prepare for greater movement in the occult. Here's the passage in full:[374]

"It is the work of the disciple to ascertain the best method for bringing

371 Filmore, Metaphysical Bible Dictionary
372 Clymer, The Science of the Soul
373 http://www.amazon.com/Deceived-Purpose-Warren-B-Smith/dp/0976349205
374 "Initiation, Human & Solar" by Alice Bailey, Ch 6, The Lodge of Masters, http://www.sacred-texts.com/eso/ihas/index.htm

about the desired results, and to formulate the plans whereby a certain percentage of success will be possible. Then he launches his scheme, founds his society or organisation, and disseminates the necessary teaching. Upon him rests the responsibility for choosing the right co-workers, for handing on the work to those best fitted, and for clothing the teaching in a presentable garb. All that the Master does is to look on with interest and sympathy at the endeavour, as long as it holds its initial high ideal and proceeds with pure altruism upon its way.

"The Master is not to blame should the disciple show lack of discrimination in the choice of co-workers, or evidence an inability to represent the truth. If he does well, and the work proceeds as desired, the Master will continue to pour His blessing upon the attempt. If he fails, or his successors turn from the original impulse, thus disseminating error of any kind, in His love and in His sympathy the Master will withdraw that blessing, withhold His energy, and thus cease from stimulating that which had better die. Forms may come and go, and the interest of the Master and His blessing pour through this or that channel; the work may proceed through one medium or another, but always the life force persists, shattering the form where it is inadequate, or utilising it when it suffices for the immediate need.

- "A very important point is now advanced. It is disciples who found the various schools of thought fostered by the energy of the Lodge. The Masters, Who inspire from behind the scenes, are not directly responsible for that which is founded. The karma of the various organizations founded falls upon the founding disciple or disciples.
- "The Masters are not yet to be found working *openly* amongst humanity. Their disciples are the forward-guard, the forerunners of the *manifest presence* of the Master later to come."

If you have an electronic version of the Bible on your iPad or smartphone, you won't find the phrase "manifest presence" in the King James Bible or the New International Version.

So how is it that Bill Johnson's Bethel Church has pioneered a global "Christian" phenomenon calling itself "Manifest Presence" – a phrase that does not appear in the Bible but which DOES appear in a 1922 Masonic Theosophy occult initiation plan for the world. Go figure. In fact, the phrase is even found on the Lucis Trust website.

From an Eastern occult text known as "The Life of the Masters of the East" comes this passage, seemingly ghostwritten by modern Pentecostal pastors:

"He saw that if man ever arose to the lofty power of that mighty Indwelling Presence; if a mighty Son of God, one in whom the Divine Wisdom abode in fullest measure; one rich in the outpouring richness of all of God's treasures, the fountain of the outpouring waters of life, the Lord, or law of compassion and wisdom; was actually to take flesh upon earth, *he must come forth and claim these possessions.*

"Then, with pure motive he must live the life and *he would bring forth that life, to which manifest Presence the name of the Christ has been given.*

"*He stood forth and boldly proclaimed that the Christ that abode in him abode in all*; that the celestial voice that proclaimed him the Beloved Son proclaimed all sons of God, joint heirs, and brothers all. This epoch is marked at his baptism when the Spirit was seen descending from heaven like a dove upon him and it abode with him. *He also said all are Gods, manifest in the flesh.*" [emphasis in the source text]

All are gods? Where have we heard that before? Remember, that's from an occultist text. It has the "name it and claim it" element common to the Pentecostal Word of Faith movement (Benny Hinn being a good example), the "manifest presence" again, and the "we are all identical to Jesus" element.

Look more carefully at the Manifest Presence line just quoted above. Lucifer followers are told that if their prayers and actions to Satan are serious enough – "with pure motive" – they will "bring forth" a "manifest Presence" which Satanists call "the Christ".

In other words, the Lucis claim is that sufficient prayers to invoke a "manifest Presence" will, in fact, conjure up the big guy with the horns and the tail.[375]

Very subtly, sharp-eyed Christians will notice that here we are not saved *through* Jesus, but *alongside* Jesus. His atoning death disappears, replaced by a doctrine of the enlightened man-god. Jesus was just an enlightened man, anointed by God, who thus becomes a god. Where have we heard that before?

This occult "sacred text" turns Jesus from saviour to merely co-saved,

---

375 Imagery of el Diablo, the Devil, Beelzebub etc has traditionally focused on the style of the Greco-Roman god Pan, half man half goat. For no other reason, probably, than convenience. Lucifer in biblical terms was an angel akin to Michael and Gabriel.

and in doing so alters the most fundamental Christian doctrine of all. No longer has he paid the price for your sin. He is merely a wise guru who has reached a higher stage of enlightenment and whose example we should follow.

If a church near you preaches occult doctrines like 'Manifest Presence' through the mouths of so-called Christian leaders, think long and hard about what they really represent. Followers who tithe into such churches are not tithing into the "Kingdom of God". They may be paying someone, but which particular Pan-Piper that is and what tune he'll play may not be to your ultimate taste.

Manifest presence is not the only Lucis Trust phrase being used in Pentecostal churches. Rodney Howard-Browne, another Bethel preacher, runs a "revival" ministry with "Great Awakening" days.[376] The phrase does not appear anywhere in the Bible, but repeatedly appears in the writings of the Lucis Trust and features in reference to the occult prayer "The Great Invocation":

"Groups worldwide ... are experiencing an answer to The Great Invocation and ... recognize through their own experience that the time of the great awakening is now!"[377]

Again, regardless of what you believe, you have to ask the question: why would supposed Christian churches be using a rallying call that doesn't appear in the Bible anywhere, but which is associated with Satanism?

How far down the rabbit hole do you want to go?

Joel Osteen runs America's biggest church, Lakewood in Houston, Texas. It's the home of a former baseball stadium, and he fills it up every Sunday. He also runs "Nights of Hope" events, and charges between US$15 and US$200 a ticket. That's revenue of at least $500,000 a night, just on the lowest ticket prices alone.[378] Joel Osteen Ministries makes around US$73 million a year in tithes and donations, based on financial data provided to *60 Minutes* in 2007. For just one of his many books, he was paid a $13 million royalty advance.

What would the Jesus who overturned the tables of the moneylenders in the temple say?

Possibly, "It is written, my house is the house of prayer, but you have made it a den of thieves!" (Luke 19:46)

---

376 https://www.revival.com/great-awakening-live/
377 "The Great Invocation", http://www.lightparty.com/Spirituality/Evocatio.html
378 http://www.youtube.com/watch?v=d_1yc_vUMC8&feature=fvwrel

Benny Hinn Ministries is another. It sells its prayer services, just like New Age practitioners. Hinn sends out letters telling people that if they donate to his ministry, no matter how hard times are for them, he will "anoint" their prayer request card and give them 12 months of prayer covering for a guaranteed return.[379]

Osteen is another of those "all paths are valid" preachers; same as the Lucis Trust.

You will recall the UN's Assistant Secretary-General Robert Muller wrote that the UN's goal of bringing back "the Coming One" had been "foretold by the great prophets". That reminder to check the prophecies is just as useful when it comes to the mega-churches like Bill Johnson's Bethel Church in California.

Bethel is a very odd name to choose for a church, if you know your Bible, because it was linked to Satan, even then:

In 1 Kings 12, Bethel is the scene of a golden calf false idol being worshipped by the masses. In 2 Kings 2, God's genuine messenger Elisha is mocked by residents of Bethel. In Jeremiah 48 we are reminded that "the House of Israel was ashamed when they trusted in Bethel."

The town has fallen from grace so far by this stage that in the Bible it is given a nickname, "Beth Aven", or "House of Wickedness".

The Lord, through Hosea, warns of the fate awaiting Bethel for promoting false teaching: (Hosea 10:1-15)

"Israel was a spreading vine;
he brought forth fruit for himself.
As his fruit increased,
he built more altars;
as his land prospered,
he adorned his sacred stones.
2 Their heart is deceitful,
and now they must bear their guilt.
The Lord will demolish their altars
and destroy their sacred stones...The people who live in Samaria fear for the calf-idol of Beth Aven. Its people will mourn over it, and so will its idolatrous priests...

"...you have planted wickedness,
you have reaped evil,

---

379 http://www.youtube.com/watch?v=jrSjylaGsn8

you have eaten the fruit of deception.
Because you have depended on your own strength
and on your many warriors,
14 the roar of battle will rise against your people,
so that all your fortresses will be devastated —
as Shalman devastated Beth Arbel on the day of battle,
when mothers were dashed to the ground with their children.
15 So will it happen to you, Bethel,
because your wickedness is great."

The Book of Amos, chapter 7, bears witness to the idolatry and heresy of Bethel, and the Lord's disgust. The arguments used by Bethel then, bear an uncanny resemblance to arguments used by Bethel today – "how dare you criticise God's anointed ones?": (Amos 7:7-17)

"7 This is what he showed me: The Lord was standing by a wall that had been built true to plumb, with a plumb line in his hand. 8 And the Lord asked me, "What do you see, Amos? "

"A plumb line, " I replied.

Then the Lord said, "Look, I am setting a plumb line among my people Israel; I will spare them no longer.

10 Then Amaziah the priest of Bethel sent a message to Jeroboam king of Israel: "Amos is raising a conspiracy against you in the very heart of Israel. The land cannot bear all his words. 11 For this is what Amos is saying:

"'Jeroboam will die by the sword,
and Israel will surely go into exile,
away from their native land.'"

12 Then Amaziah said to Amos, "Get out, you seer! Go back to the land of Judah. Earn your bread there and do your prophesying there. 13 Don't prophesy anymore at Bethel, because this is the king's sanctuary and the temple of the kingdom. "

14 Amos answered Amaziah, "I was neither a prophet nor the son of a prophet, but I was a shepherd, and I also took care of sycamore-fig trees.

15 But the Lord took me from tending the flock and said to me, 'Go, prophesy to my people Israel.' 16 Now then, hear the word of the Lord. You say, 'Do not prophesy against Israel, and stop preaching against the descendants of Isaac.'

17 "Therefore this is what the Lord says:

"'Your wife will become a prostitute in the city,
and your sons and daughters will fall by the sword.

Your land will be measured and divided up,
and you yourself will die in a pagan country.
And Israel will surely go into exile,
away from their native land. '"

This, then, is the fate of Bethel. Apart from a similar forlorn and fleeting reference in Zechariah 7, the town is never mentioned again in the Bible. What had begun as a city whose name meant "House of the Lord", with such promise, ended in heresy and arrogant human pride, abandoned by God himself who left Bethel to its delusions and passed judgement accordingly.

Again, you couldn't make this stuff up: One of America's biggest megachurches turns out to be preaching New Age concepts and names itself after a Godforsaken false temple in the Bible. What genuine Christian pastor would name his church after something nicknamed "The House of Wickedness"? It's another of those bizarre cognitive dissonance moments like pastors John Lake and Rick Joyner endorsing the "Knights of the Morning Star", which translates literally as "Knights of Lucifer".

Yet thousands of Americans and others from around the world are paying this Bethel church maybe a million dollars each week, its influence on worldwide Christianity is huge, and it's as if its followers have been blinded to what the Manifest Presence movement of Bethel is really about.

There is another fundamental clue parallel between the Luciferian doctrine and modern Christian evangelism of the past century: both are preaching Kingdom-now theology. That is, both groups say that if consciousness is raised and there is a "revival", we can usher in the Kingdom of God.

There is barely a Pentecostal church in the West that is not following this doctrine in some way shape or form. There's only one glaring problem with that – no such "revival" appears in Biblical prophesy whatsoever. In fact, the word "revival" does not actually appear in the Bible. Not in the NIV, not in the King James Version. The concept, that the world will turn to Christ in a major "revival" and usher in his Kingdom on Earth, does not appear in the Bible either. Yet go into almost any large evangelical church these days, and this concept is being preached.

This is not to confuse the issue with minor "revivals", of which there have been many through the centuries as the church goes through growth periods. But always, like a moat full of crocodiles, no matter how big any revival is, the prophecies say faith peters out and there is a period of utter persecution of the Church prior to the return of Christ, and not a sniff of a revival anywhere.

Therefore, extrapolating this idea out to a specific belief that such a "revival" on a global scale will usher in the return of Christ is simply not biblical. It is, however, exactly what the Luciferians claim, except they call it "the Coming One".

This is another huge clue that Christian churches have been conned into preaching New Age doctrine.

In the biblical texts, if you study them, Jesus clearly warns that things will get increasingly hard for the Christian church as the end nears. More and more people will fall away from the faith, rather than be revived into it. Let it sink in again: there will be no major revival that coincides with the return of Jesus Christ. It is not prophesied in Revelations, nor by Jesus directly. Quite the opposite in fact.

It is written:

"Enter through the narrow gate. For wide is the gate and broad is the road that leads to destruction, and many enter through it," Jesus said. "But small is the gate and narrow the road that leads to life, and only a few find it." As if predicting the exact scenario we are debating in this book, Jesus then immediately adds, "Watch out for false prophets. They come to you in sheep's clothing, but inwardly they are ferocious wolves. By their fruit you will recognise them." (Matt 7:13-16)

No matter how many ways you try to slice and dice that, the "narrow" path that "few" will find is a far cry from the mass revival ushering in the Kingdom that many pastors are preaching today. From a Bible perspective, it is actually impossible to reconcile many modern Christian "doctrines" with the Bible.

This new concept of Kingdom Now theology is also known as "Dominionism". It sounds Christian, but taken too far it becomes anti-biblical. The entire Bible, you see, is filled with end-time prophecies. Regardless of whether one believes them or not, whether one is atheist or Christian, it is a plain simple fact that the Bible predicts very dark times to come for true believers, and therefore a theology that preaches something else like fluffy rabbits and happy times is by definition a false gospel.

Dominionists believe the earth needs to be ruled righteously in readiness for the return of the Lord, and they believe the Church is the instrument through which this ushering in of the Kingdom will be achieved.

Luciferians believe the earth needs to be ruled righteously in readiness for the return of the Coming One, and they believe the United Nations is the instrument through which this ushering in of the Kingdom will be achieved.

You can begin to see the problem here. The churches I'm sure will try to argue the pagans are copying. The evidence suggests it's the other way around.

When Jesus spoke of his Kingdom in the New Testament, it was discussed primarily at two levels. The first was the kingdom of God breaking into the kingdom of Satan's realm (the world), which began at the cross. From that moment forward, every baptism took one more soul away from the potential clutches of Hell and therefore is said to have gained spiritual ground for God in the natural world.

For those of you who don't have a religious belief, please try and follow the claims in the passages that follow, because the point is this: they will prove to you that a massive plank of the modern "Christian" church is in fact directly contrary to what the Bible says, and that should be making your ears prick up.

The *Evangelical Dictionary of Theology* expresses this point on the real Kingdom of God and its implications so clearly that I'll quote it directly:[380]

"People may sow the seed by preaching the kingdom (Matt 10:7; Luke 10:9; Acts 8:12; 28:23,31); they can persuade others concerning the kingdom (Acts 19:8), *but they cannot build it.*"

Read those last few words again. Churches can sow the seeds that lead to salvation which in turn builds the kingdom, but they cannot build the kingdom themselves. No amount of "revival" is going to "bring it on".

"It is God's deed," continues the passage. "People can receive the kingdom (Mark 10:15; Luke 18:17), but they are never said to establish it. People can reject the kingdom and refuse to receive it or enter it (Matt 23:13), but they cannot destroy it. They can wait for it (Luke 23:51), pray for its coming (Matt 6:10), and seek it (Matt 6:33), but they cannot bring it.

"The kingdom is altogether God's deed although it works in and through humans. People may do things for the sake of the kingdom (Matt 19:12; Luke 18:29), work for it (Col. 4:11), suffer for it (2 Thess. 1:5) but they are not said to act upon the kingdom itself. They can inherit it (Matt 25:34); 1 Cor. 6:9-10, 15:50), but they cannot bestow it upon others."

In other words, any church that says it has been called by God to

---

380 Evangelical Dictionary of Theology, ed. by Walter Elwell, Baker Academic, 2001, p658

establish his Kingdom on earth has misunderstood both its power and its mission. Its job is simply to preach the gospel, sowing the seed. According to the Biblical texts, the harvest is God's work. By praying for others in healing and deliverance, Christians assist the kingdom of God, but they do not "bring it", "establish it" or "bestow it". They are not given that authority – and that is a key point.

Yet you have now seen the Lucis Trust writings. Both Lucis and the modern evangelical church claim that by raising awareness, they can manifest the Presence now.

Given that you now know "manifest presence" does not appear in the Bible anywhere, but that it *is* an occult term associated with calling in the Coming One, the deadly seriousness of what you are reading here should be sinking in. We saw in an earlier chapter that the same huge investment banks and funds controlled all the global food and chemical companies – they were different brands but effectively controlled by the same people. Is the Lucis Trust likewise the real controller of religious belief in the West and Asia? You can worship any brand you like, but the core message is the same.

A moment ago we discussed there being two ways of looking at the Kingdom. Let's examine the second. While a personal manifestation of the kingdom is established with the conversion of each new believer, as just shown, Jesus made it clear that his kingdom would not be properly established until his return at the end of the age. "Your kingdom come," he prayed in the Lord's Prayer, not "your kingdom now". For now, the kingdom grows alongside that of "the prince of this world" (Jn. 14:30)

How has it come to this? Why are churches all over the world following teachings that are not found in the New Testament? How the heck did something so monumentally big slip past the goalposts without being noticed?

The UN's Robert Muller urged people to read the prophets. Well, in the Book of Ezekiel, God warns that his judgement will "begin at my sanctuary", the very temple itself.[381] Are people who follow these charismatic preachers taking the broad "road that leads to destruction" that Christ warned of?

None of this is recited to convert you one way or the other. In a book about an occult plot to infiltrate churches and global organisations, it

---

381 Ezekiel 9:6

is necessary to compare common features and differences, in order to understand what is playing out.

Some of you may be arguing at this point, and rightly so, that your local church still does baptisms. This may be so, but if they are preaching concepts that do not appear in the Bible, but which do appear in the occult texts, *what exactly are people being baptised into?*

There's that intriguing and difficult prophecy from Jesus Christ himself who warned a time would come, at the end, when those professing to be Christian would be turned away:[382]

"Not everyone who says to me, 'Lord, Lord,' will enter the kingdom of heaven, but only the one who does the will of my Father who is in heaven. [22] Many will say to me on that day, 'Lord, Lord, did we not prophesy in your name and in your name drive out demons and in your name perform many miracles?' [23] Then I will tell them plainly, 'I never knew you. Away from me, you evildoers!'"

I cannot put it plainer than this: the fox is in the henhouse. Modern Pentecostal worship has not merely deviated from the path; it's run completely off the road and down a goat track. And not just any old goat track. This one is named Lucifer Road, and the old goat who lives at the end of it must be laughing his horns off. The Beast is in a pulpit, somewhere near you.

Millions of Christians, genuine people, led astray by slick preachers, stirring music, and sermons that push all the right buttons but are not biblically accurate. Sermons and "inspirational" books and DVDs whose contents originated in Satanism. Possibly the biggest deception in the history of the Church.

Little wonder hundreds of millions of people in the West have drifted away from the Christian faith: the modern churches are a cardboard cut-out, fleecing the few faithful who remain while now preaching an almost identical spiritual doctrine to those who worship Lucifer. They may use the name 'Jesus' in their services, but the phraseology and the doctrine is essentially Lucis Trust material.

The New Age takeover of the United Nations you could put down as a one-off. The head of Roman Catholicism, Pope Paul VI, bending his knee at an altar to the Devil then handing his papal ring to the head of the UN and swearing allegiance to the UN is something out of a B grade horror

---

382 Matt. 7:21-23

movie like *Invasion Of The Body Snatchers*. To then find the Lucis Trust's occult phrases and concepts being preached from prosperity-doctrine, "Word of Faith" Pentecostal mega-churches and 'Christian' radio and TV networks across the world and in books sold in every "Christian" bookshop makes it utterly impossible to explain. This is a global phenomenon; how did it happen? If you have a rational explanation as to how these events are all examples of coincidence theory, by all means share it.

And you've got to hand it to the Satanists – not only have they knocked out most of the churches, they've also captured Buddhism, as we saw earlier in the book. The strains of Buddhism practiced today across Asia and the West stem from a revival in the late 1800s and early 1900s directly caused by the association of revivalist Buddhists like D T Suzuki – the father of the western Zen movement – with Helena Blavatsky's Theosophy teachings on Lucifer.[383] [384]

As Buddhism scholar Robert Sharf noted, the few remaining authentic Buddhist strains in the East regard the modern popular Buddhism with contempt.[385] So if you are a western or modern Asian Buddhist reading this, laughing at the misfortune of the Catholic Church and the happy-clappy Pentecostals, congratulations, it turns out you've been 'punked' as well.

Suddenly, the discovery that the alleged birthplace of "Lord Buddha" at Lumbini carries the official designation 666 in the UNESCO World Heritage system makes eerie sense. Buddhism, as we know it today in its Zen, Theravada and Mahayana varieties is like the façade of an old heritage building hiding something modern behind it. The modern bit is Lucis Trust Satanism to the core. The façade is the shadow of what Buddhism used to be.[386]

---

383 Algeo, Adele S. "Beatrice Lane Suzuki: An American Theosophist in Japan." Quest 95.1 (JANUARY-FEBRUARY 2007): 13-17. http://www.theosophical.org/publications/quest-magazine/1254

384 Satanists Helena Blavatsky and Henry Olcott visited India and Sri Lanka in the late 1870s to spread Theosophy among the Buddhist schools of the East. Theravada, Mahayana and Zen are among the most infested, as is the Vipassana variant. http://www.theosophical.org/henry-s-olcott/his-work-with-buddhism

385 As Sharf writes: "Suffice it to say that, just as the writings of Suzuki and Hisamatsu are not representative of traditional (i.e., pre-Meiji) Zen exegetics, the style of Zen training most familiar to Western Zen practitioners can be traced to relatively recent and sociologically marginal Japanese lay movements which have neither the sanction nor the respect of the modern Rinzai or Sōtō monastic orthodoxies." See "The Zen of Japanese Nationalism" by Robert H. Sharf *History of Religions*, Vol. 33, No. 1. (Aug., 1993), pp. 1-43. The University of Chicago Press, http://www.thezensite.com/ZenEssays/HistoricalZen/Zen_of_Japanese_Nationalism.html

386 Buddhism had become virtually extinct in India for centuries, but was revived when the Blavatsky-inspired Buddhists in Sri Lanka went back to the mainland to open new schools of training.

It's no longer like merely watching a B-grade horror, it's almost as if we are trapped in one. Now is not the time to panic, however, because this occultism has spread even further and it has its paws directly on the global governance agenda.

You may not accept any of the supernatural beliefs from any of the protagonists in this story, but as I have said before, what you believe is irrelevant.

If you are a rationalist or an atheist and you've been focusing your attention on Christianity in society, then with respect you were successfully distracted. The impact of the New Age on your life if the totalitariacrats put their new system in place will far outweigh any Christian impact on your daily life. There's a good chance you'll look back one day and long for a return to the current state of affairs.

The UN's New Age change team is winning on virtually every major political front, and the two main Christian denominations have largely fallen, as has Buddhism. The spiritual beliefs of somewhere between 1.5 and 3 billion people are effectively in the hands of the Church of Lucifer. Therefore, what *these people* believe *will* affect you, whether you like it or not. If this movement is strong enough to make a Pope fold, it would be foolish to assume you are home free.

These people also have access to some very serious money. Archived tax haven bank records suggest that a "Windsor International Bank & Trust Company" based on the Caribbean island of Grenada was used to funnel funds to the Lucis Trust and a range of other key players. The list, an official disclosure statement filed with the Grenada authorities in 2000, makes intriguing reading:

"The Bank is also a Member of, Advisor to, Affiliate of or Contributor to the following Organisations:

- International Fund For Development
- The Hall Family Foundations
- The Rockefeller Foundation
- WHO/Habitat For Humanity
- The Lucis Trust (NGO; United Nations)

Sri Anagarika Dharmapala was one of Blavatsky's key disciples and instrumental in bringing the satanic form of Buddhism to dominance in the 20th century. He acted as a translator on Henry Olcott's theosophical works. Decades later he split with the Theosophical Society over their plans for a universal religion, but by then the horse had long since bolted. Buddhism would never be the same.

- National Resources Defence Council
- Capital Missions Company
- Investors Circle
- The Coca-Cola Foundation
- Fellowship For International Education
- International Monetary Agency
- International Center For Educational Advancement
- Christian Children's Fund (Worldwide)
- Fellowship For Reconciliation
- National Institute For The Advancement Of Science
- International Association For Environmental Cooperation
- World Wildlife Federation
- International Trading Alliance
- Council For Foreign Relations
- CARICOM
- NAFTA
- MIRCOSUR
- Council Of Emerging Nations
- Freedom Communications, Inc.
- The European Institute
- United Nations Association Of The USA

There are some massively big names in there, and an obscure West Indies tax haven bank was, on the face of it, the entity used by somebody to get money to them.[387]

Who would use secret tax havens to fund Satan worshippers and other Green organisations? Clearly someone with a lot of money...

---

387 The Windsor International Bank & Trust is now defunct, but at the time its directors included two Grenadian magistrates and a former Prime Minister. For the official version of the above list, See http://web.archive.org/web/20001018174144/http://www.windsorbank.com/contact.html#notice

CHAPTER 23

The Devil Made Me Do It

> "The Matrix is everywhere. It is all around us. Even now, in this very room. You can see it when you look out your window or when you turn on your television. You can feel it when you go to work... when you go to church... when you pay your taxes. It is the world that has been pulled over your eyes to blind you from the truth."
>
> *– Morpheus, The Matrix*

It's fair to say the United Nations has, almost since its inception, been a hotbed of New Age belief. Dr Robert Muller was a key player in devising the structure and ethos of the UN Development Programme[388], currently run by former New Zealand Prime Minster Helen Clark. He was instrumental in UNESCO, the United Nations Educational, Social and Cultural Organisation – an ideal platform from which to spread the occult belief system – and he was also heavily involved in setting up the population control agency, the UN Population Fund, and the World Youth Assembly.

You might think UNESCO's role was supposed to be charitable work with those in need. It wasn't. As Muller's co-founder Julian Huxley explained in UNESCO's mission documents, it was to prepare a global brainwashing programme via schools:[389]

---

388 As acknowledged by UN Secretary General Ban Ki-moon at Muller's eulogy in 2011
389 http://unesdoc.unesco.org/images/0006/000681/068197eo.pdf

"Further, *since the world to-day is in process of becoming one, and since a major aim of UNESCO must be to help in the speedy and satisfactory realisation of this process*, that *UNESCO must* pay special attention to international education – to *use education as a function of a world society*, in addition to its functions in relation to national societies, to regional or religious or intellectual groups, or to local communities."[390]

UNESCO would not just attempt to influence national school curricula, it would use friends in the news media to push the agenda as well:

"There are thus two tasks for the Mass Media division of UNESCO, the one general, the other special. The special one is to enlist the press and the radio and the cinema to the fullest extent in the service of formal and adult education, of science and learning, of art and culture. The general one is to see that these agencies are used both to contribute to mutual comprehension between different nations and cultures, and also to promote the growth of a common outlook shared by all nations and cultures."[391]

The Lucis Trust interviewed Erskine Childers, a Senior Advisor to the UN Director-General for Development and International Economic Cooperation. Childers said the most important policy to help change the world was raising "awareness" of the United Nations in all schools.

"[People] can undertake to increase knowledge about the about the UN amongst some identified group of their fellow citizens. Don't pick too ambitious an area to start with or you might find that unmanageable. Take a shire or a county. You might find out how the UN is taught in schools in the area.

"Just asking questions about teaching on the UN in schools can bring about change. An academic in Oxford asked me, 'Are you suggesting that the UN ought to be part of the curriculum'? I was amazed. 'My God,' I said, 'isn't it part of the curriculum now?'"[392]

In an interview with the Lucis Trust published on the 50th anniversary of the United Nations in 1995, Robert Muller remarked that his former boss, Secretary-General U Thant, had once described the establishment of the UN as the most important event in the 20th century. Bigger than World War II, bigger than Ben Hur. Muller told the Lucis Trust that "the World Commission on Education in the 21st Century, established by UNESCO, is the second."[393]

390 Ibid p29
391 Ibid p60
392 World Goodwill newsletter 1995, http://www.lucistrust.org/en/content/download/1722/21149/file/1995-2.pdf
393 Ibid

If you have any lingering doubts that Muller's influence on global education policy is major, that quote should lay those doubts to rest. Muller was endorsing UNESCO's education curriculum as a crucial development because it is exactly what he wanted.

He wasn't interested in ordinary schooling; go back and read his writings in depth. His overwhelming doctrine was raising planetary spiritual consciousness by working the sacredness of nature and the special mission of the UN into the school curriculum.

You will find heavy emphasis on the supremacy and wisdom of the United Nations in the school curricula of Australia[394], New Zealand[395], and Britain[396]. In Canada education is run by each of the states, but British Columbia as a random example is locked in with the UN agenda, as a list of classroom teaching resources shows.[397] We will get to the United States shortly, with its Common Core education plan.

Another big project of Robert Muller's was the United Nations University of Peace, built on a mountain in Costa Rica of all places. Why? Because Muller believed he was fulfilling an ancient Indian prophecy:[398]

"Dear children, the Great Spirit is in every animal, in every bird, butterfly, flower, insect, leaf and grass you see. The Great Spirit is also in you, the Creator's children. Please take care of the wonderful nature created by God and some day, from this mountain, you will see the birth of a civilization of peace spread to the entire world."

If you think that the UN would not waste aid donations on one person's spiritual quest, think again. An entire university was built to fulfil a prophecy.

One of Muller's fellow travellers at the UN was a man named Maurice Strong. Born in 1929 in Canada, Strong cut his teeth in the oil industry but maintained close ties with UN staff in the US and Canada and quickly fell, at the age of 18, into the orbit of change agent and business mogul David Rockefeller. In 1971 Strong commissioned an environmental report examining the state of the earth's ecosystems, and tabled the document,

---

394 http://www.australiancurriculum.edu.au/Search?q=%22united+nations%22
395 http://www.tki.org.nz/tki-content/search?SearchText="united+nations"
396 http://www.education.gov.uk/search/results?q=%22united+nations%22
397 http://www.bced.gov.bc.ca/irp/pdfs/social_studies/support_materials/sj12_websites.pdf
398 Well, that's Muller's claim. He calls it the "Mt Rasur Legend" on his website, but doesn't disclose the "legend" was actually a poem published in 1946 by a Costa Rican politician named Roberto Brenes Mesén. The "legend" was younger than Muller himself by some decades and the writer of the poem was a Lucifer worshipper and a freemason. It's what you might call a "self-fulfilling prophecy". http://www.robertmuller.org/mr/p01.html

eventually published as a book, at the world's first environment conference in Stockholm in 1972.

It had come about like this. The Socialist government in Sweden were on good terms with UN officials given that one of their own had been UN Secretary-General for a time, and in 1968 they suggested to the UN that they could host a world environment conference if the United Nations was prepared to put some funding and effort into it. The then-Secretary-General U Thant and Robert Muller tapped Maurice Strong on the shoulder and asked him to organise the project.

Some 400 delegates from 113 countries attended, the end result of which was the establishment of United Nations Earth Day, and a commitment by delegates to change the world, one step at a time:[399]

"The Conference could not deal with all the ills of the world, but if it successfully accomplished the important work before it, it would establish a new and more hopeful basis for resolving the seemingly intractable problems that divided mankind. It had to be recognized that the physical interdependence of all people required new dimensions of economic, social and political interdependence."

Maurice Strong, in an interview with the BBC on the eve of the Stockholm conference, displays hints in his language that he was already walking on the dark side of the force:[400]

"Whether we are pessimistic or optimistic depends on the nature of man. Whether man, in the light of this evidence, is going to be enlightened enough and wise enough to subject himself to this kind of discipline and control," Strong said on British TV.

"Stockholm must usher in a new era of international cooperation, and I believe that this will be the dominant motivation."

Strong also told the BBC that countries like Canada might have to issue licences restricting the number of children in a family. Of all the countries in the world with overpopulation problems, vast sprawling Canada wasn't the one that first came to mind.[401]

Strong's prominence and leadership[402] saw him appointed by the UN to run the newly-established United Nations Environmental Program (UNEP),

---

399 "Brief summary of General Debate, Stockholm Conference 1972, UNEP website, http://www.unep.org/Documents.Multilingual/Default.asp?DocumentID=97&ArticleID=1497&l=en
400 http://www.youtube.com/watch?v=1YCatox0Lxo
401 In 1970, with a population of 21 million spread across 10 million square kilometres, Canada's population density worked out at one person per 50 hectares (125 acres)
402 Maurice Strong also joined Robert Muller's "University of Peace" as its President.

which would plan for future environment protection globally. Their efforts led UN Secretary-General Javier Perez de Cuellar to establish, in 1983, the World Commission on Environment and Development (WCED).

This organisation's name was quickly shortened in the media to "the Brundtland Commission" after the appointment of Norway's socialist leader and first woman prime minister, Gro Harlem Brundtland, to lead the inquiry. Also on the Commission were Maurice Strong and William Ruckelshaus.

Brundtland was a key figure in the Marxist UN advisory group, Socialist International.[403] Ruckelshaus was a former Nixon appointee who'd seen the "light" and become the first Director of the US Environmental Protection Agency. Among other things, Ruckelshaus has been on the board of directors of global chemical giant and genetically modified food producer Monsanto. He was also director of a water rights exploitation company established by Maurice Strong to sell water from a New Age retreat on a property Strong owns at Crestone, Colorado.

There were other international representatives (most with a UN connection) on the Brundtland Commission, perhaps the most notable being Commonwealth Secretary-General Sir Sridath Ramphal (Sonny Ramphal to his friends). Ramphal went on later to lead a team calling for Global Governance in a major UN report.

The Brundtland Commission issued its report on global sustainability in 1987, feeding into a United Nations global governance agenda that was already well underway. In 1988 climate activists James Hansen and Tim Wirth cranked up the heat in a US senate hearing room the night before they testified about the relatively new concept of global warming. Tim Wirth, a Democrat Senator who became head of the United Nations Foundation later told an interviewer they did it as a PR stunt to get the politicians on side:[404]

"Believe it or not, we called the Weather Bureau and found out what historically was the hottest day of the summer. Well, it was June 6 or June 9 or whatever it was, so we scheduled the hearing that day, and bingo: It was the hottest day on record in Washington, or close to it.

"What we did was went in the night before and opened all the windows,

403 In a 2005 report to the UN, Socialist International and the UNDP advocate the need to establish a world government. Read the full story and get the link to the UN report here: http://www.investigatemagazine.co.nz/Investigate/2559/global-governance-on-climate-agenda/
404 PBS interview with Timothy Wirth, 17 January 2007, http://www.pbs.org/wgbh/pages/frontline/hotpolitics/interviews/wirth.html

I will admit, right? So that the air conditioning wasn't working inside the room and so when the, when the hearing occurred there was not only bliss, which is television cameras in double figures, but it was really hot. … So Hansen's giving this testimony, you've got these television cameras back there heating up the room, and the air conditioning in the room didn't appear to work. So it was sort of a perfect collection of events that happened that day, with the wonderful Jim Hansen, who was wiping his brow at the witness table and giving this remarkable testimony."[405]

In a speech at Cornell University, Wirth told students that the billion dollar United Nations Foundation established by media magnate Ted Turner, and run by Wirth, is a huge believer in "public private partnerships" to solve global problems. "It is our challenge to figure out a fair and equitable way to share our global commons, such as air and water."[406]

You can take that as code for "commercialisation" of the air you breathe and the water you drink, through public/private partnerships.

It's a theme utterly familiar inside the whole climate change/environmental movement. The public are duped through simplistic messages, chants and media manipulation into going along with a plan that ultimately hands power and economic control to a global business and intellectual elite. While common or garden-variety New Age speakers have figured out how to sting their gullible public followers $1500 each for a seminar, a photo-op and a free book, people at the absolute top of the food chain have figured out how to do it at a global level, using their skills to get taxpayers' cash by the billions.

What we are seeing unfold is that "convergence of opportunity" mentioned at the start of this book. There are a huge section of top activists motivated primarily by power and a spiritual element. Then there are those who realised that jumping on the bandwagon allowed them to make a lot of money. These people don't care who is in "power" as long as they control the process.

It's a variation on a famous merchant banker's quote.

"Give me control of a nation's money supply, and I care not who makes the laws," boasted Meyer Amschel Rothschild, the founder of the banking dynasty, back in the late 1700s.

---

405 A similar stunt was pulled with the timing of the Earth Summit at Rio in 1992, carefully arranged to coincide with historically the hottest day in tropical Brazil.
406 "Effective climate change strategies call for new rules in global politics and economics", *Cornell Chronicle*, 21 April 2008, http://www.news.cornell.edu/stories/2008/04/wirth-calls-major-changes-global-politics

One of the mechanisms the globalists are hoping to introduce are "carbon accounts" for every man woman and child on the planet, and making them equal. As the *Cornell Chronicle* reported, Wirth "suggested cutting carbon emissions so that every individual would be allotted the same amount. We're far from equal allotments now, Wirth explained, with each U.S. citizen emitting 20 tons of carbon into the atmosphere each year, while the average is 11 tons per person in Germany, 3 tons per person in China, and 1 ton per person in India."

Tim Wirth is a globalist, and he's not averse to business opportunities arising from that. He was one of the instigators – along with disgraced Enron boss Ken Lay – of the crafty plan to buy and sell "carbon credits". Wirth was committed to forcing the world to adopt cap and trade and other commercial climate change measures, regardless of whether climate change is actually happening:

"We've got to ride the global warming issue," said Wirth. "Even if the theory of global warming is wrong, we will be doing the right thing in terms of economic policy and environmental policy."[407]

It didn't actually matter that carbon trading was of no net-benefit to the planet (simply transferring emissions rights from one country to another), because it provided a new trading mechanism for ticket-clipping financiers on the world market. In reality, emissions trading schemes based on carbon markets were nothing more than an extension of socialist ideology: wealth transference from the taxpayers and consumers of the first world, to the governments and leaders of the third world.

In a profile of Wirth's colleague Maurice Strong, Australia's *Quadrant* magazine made a similar point: there's a lot of money to be made and power to be gained in the business of administering climate change laws:

"[Strong's] private interests always seemed to be in conflict with his public persona and his work on the world stage. Strong's extensive range of contacts within the power-brokers of the world was exceptional. One admirer christened him 'the Michelangelo of networking'. Maurice Strong described himself as 'a socialist in ideology, a capitalist in methodology'."[408]

Maurice Strong, however, was not just a businessman, and not just a socialist. He was also a devout New Ager[409]. The actual term "New

---

407 National Review, January 27, 1997, page 10.
408 "Discovering Maurice Strong" by John Izzard, Quadrant, 31 January 2010
409 Technically, a Ba'hai, like Gorbachev and Al Gore

Age"[410] was given extra marketing muscle by Maurice Strong's friend, David Spangler, and Spangler has defined the implications of the New Age, saying that everyone left on planet Earth will eventually have to submit to a "Luciferic initiation":

"Lucifer comes to give us the final gift of wholeness. If we accept it then he is free and we are free. That is the Luciferic initiation. It is one that many people now, and in the days ahead, will be facing, for it is an initiation into the New Age."[411]

Vampire bat or fruit bat? Either way, it's the same old guano: the totalitariacrats include people who genuinely believe humanity must swear allegiance to the Devil, and that in the act of swearing allegiance, Lucifer is set free. The way Spangler describes it, Satan sounds like a dirty old man in a raincoat offering sweeties to children:

"Lucifer becomes something else again. He becomes the being who carries that great treat, the ultimate treat, the light of wisdom," says Spangler in what is effectively a repeat of the deal first offered in the Genesis story of the Garden of Eden – "eat of this fruit and ye shall become as gods".[412]

One who has done so is Maurice Strong. The Canadian billionaire purchased, in the mid 1980s, a multi-thousand acre property in the Colorado mountains known as "Baca". Some reports say it is 63,000 acres in size, others up to 200,000 acres. The latter is probably correct – it's contained in a news article Maurice Strong posted on one of his websites, and interviews given by his wife.[413] The picture is a little confusing because Strong appears to have sold or given away some of the land to help create New Age Grand Central – America's pilgrimage site for the New Age. The site is administered by the Strong family's Manitou Foundation which in turn was funded by the Rockefeller family to the tune of US$100,000 a year.

The land had originally belonged to Saudi arms dealer Adnan Khashoggi, who in turn flicked it on to Strong. In the news article, based on an interview with Strong's second wife, Hanne, we learn she fell in love with the remote landscape – a mountain range named Sangre de Cristos, or "Blood of Christ" – as a spiritual retreat:

---

410 The phrase first appears in any meaningful sense in Helena Blavatsky's ode to Lucifer, "The Secret Doctrine" published in 1888. The Lucis Trust's Alice Bailey hit the home run with the book "Discipleship in the New Age" published in 1944. The term was heavily popularised by David Spangler's Findhorn Foundation in the sixties and seventies.
411 *Reflections on the Christ* by David Spangler, Findhorn Publications, 1977, p. 45
412 Ibid p41
413 http://www.manitou.org/MF/articles.php

"The Hopi had used it only for sacred ceremonies. Three months after her arrival in 1978, as she tells it a wild-haired 80-year-old named Glenn Anderson, dubbed 'the Prophet' by the locals, knocked on her ranch house door with the words: 'So you've finally come'. He proceeded to spell out a vision he had received, she says, that a woman like her would preserve all the world's faiths in the valley against some imminent doomsday."

Imagine Strong's surprise when he discovered his land sat on top of reportedly the largest freshwater reserves in North America. Hanne Strong was not the only one having a mystical experience in their new valley however.

Maurice Strong claims to have had a Moses-like experience with TV journalist Bill Moyers:

"We'd been walking, talking, heading back to my parked car. Suddenly, this bush – some sage-brush – erupted in flames in front of us! I was astounded. Moyers was, too. A bush bursting into flames... It is the most mystical experience he has had."[414]

Strong's New Age beliefs, already evident from his 1972 BBC interview, were in full swing. Having served on the socialist Brundtland Commission whose 1987 report deemed the environment and "sustainable development" as the core issues to concentrate on, Strong moved into high gear. Plans were drafted for a major world conference in Rio de Janeiro, Brazil, in 1992, to which every world leader would be invited. Maurice Strong was, again, given the task of organising it by the United Nations. To do that, he needed to raise the profile of climate change and environmentalism.

In his position as a foundation director of the secretive World Economic Forum at Davos each year, strong gave an astounding interview with *West* magazine in 1990, where he talked of collapsing the western industrial system if he had to:

"Each year the World Economic Forum convenes in Davos, Switzerland. Over a thousand CEOs, prime ministers, finance ministers, and leading academics gather in February to attend meetings and set the economic agendas for the year ahead. What if a small group of these word leaders were to conclude that the principle risk to the earth comes from the actions of the rich countries?" posed Strong.[415]

"And if the world is to survive, those rich countries would have to sign an agreement reducing their impact on the environment. Will they do it?

414 Interview with Maurice Strong, by Daniel Woods, West magazine, 1990
415 Interview with Maurice Strong, by Daniel Wood, *West* magazine, May 1990

Will the rich countries agree to reduce their impact on the environment? Will they agree to save the earth?

"The group's conclusion is 'no'. The rich countries won't do it. They won't change. So, in order to save the planet, the group decides: isn't the only hope for the planet that the industrialized civilizations collapse? Isn't it our responsibility to bring that about?

"*This group of world leaders form a secret society to bring about a world collapse*. It's February. They're all at Davos. These aren't terrorists – they're world leaders. They have positioned themselves in the world's commodity and stock markets. They've engineered, using their access to stock exchanges, and computers, and gold supplies, a panic. Then they prevent the markets from closing. They jam the gears. They have mercenaries who hold the rest of the world leaders at Davos as hostage. The markets can't close. The rich countries...?"

The journalist interviewing him describes Strong's body language as he leaves the question hanging: "And Strong makes a slight motion with his fingers as if he were flicking a cigarette butt out of the window. I sat there spellbound.... He is, in fact, co-chairman of the Council of the World Economic Forum. He sits at the fulcrum of power. He is in a position to do it."

And arguably, that's precisely what he and his followers began to do: regulate the West into economic thrall through climate change laws. Driven by spirit guides, yet with the political and economic power – still – to force change, Maurice Strong is the epitome of the coming totalitarian regime.

Journalist Daniel Wood, who penned the article, spent a week with the Strongs at their Baca mystic complex, "wondering what dedication, what idealism compels them toward such an unlikely dream. And the more I learn, the more aware I become that I've entered a world of illusions, where the surface conceals things unfathomable."

Canadian author and left-winger Jeff Wells understands the Zeitgeist. Like others, he too noted in a 2005 background piece the influence of the satanic Lucis Trust on Maurice Strong, and he repeats the point made at the start of this book: this is not a battle between left and right anymore, this is a battle for spiritual and economic control of the world. The acolytes and the elite will share the spoils, the rest of us will be the drones.

"I have to say, this is a curious place for me to be," remarks Wells. "Because much of what Strong espouses, I could say I support. I'm a socialist, I'm an environmentalist, I'm ecumenical in spirit. But I think to get at the heart of the mystery of our age, we need to take off our left or right blinders, and quit gaming our parapolitics to match our politics.

"The Canadian Strong is an oxymoronic billionaire socialist who serves as Special Advisor to the Secretary-General of the United Nations and as Senior Advisor to the President of the World Bank. He is also, among many other things, Chairman of Strovest Holdings, Chairman and Director of Technology Development Corp, Director of the Foundation Board of the World Economic Forum 'Davos' and Chairman of the Earth Council. Strong's former appointments include Secretary General of the World Bank and Director of the Rockefeller Foundation. He is a leading Bilderberger, and member of the Trilateral Commission, the Council on Foreign Relations and Club of Rome...linked to the Lucis Trust.

"Strong wasn't born to privilege, but was cultivated by David Rockefeller whom he met at 18.

"Certainly the neoconservatives are important players in the drama. (Lest we forget, Strong will now be advising Paul Wolfowitz in the latter's new role as President of the World Bank). But the neocons may themselves have been played, and too blinded by their imperial overreach to know they've become useful idiots of another agenda which will mean America's ruin. An agenda which intends to reboot the world.

"I've said it before, but there's something about Luciferians that bears repeating: they think they're the good guys," warns Wells.[416]

The phrase "useful idiots" was first coined by either Vladimir Lenin or Josef Stalin to describe Western sympathisers of communism; people who supported Marxism because it was 'cool' or 'rebellious', not because they genuinely understood the implications. It can be accurately applied to those tolerant of New Age beliefs without genuinely understanding the differences between religions. It can certainly be applied to those people who are New Age followers who don't understand the link to Satanism.[417] If you don't know you are being played – and it's within your power to know if you had made a little more effort – you are a *putz*, and that applies to conservatives and liberals alike.

On the eve of the 1992 Earth Summit at Rio, Maurice Strong's wife

---

416 "Some people call me Maurice" by Jeff Wells, 21 April 2005, http://rigorousintuition.blogspot.
co.nz/2005/04/some-people-call-me-maurice.html
417 As I've said before, if you want a more extensive theological explanation about comparative religion and why the New Age is Satanism in drag, regardless of how they choose to define it themselves, see *The Divinity Code* by Ian Wishart, Howling At The Moon Publishing, 2007

Hanne invited followers to an event, the Sacred Earth Gathering:

"Evening. The gate to Vera's ranch is opened by a watchman. We drive past gardens, corrals, dovecotes, and a schoolhouse. The ranch is built on the ruins of an early Benedictine monastery on a steep hill overlooking Rio. Hanne Strong, wife of the Secretary-General of UNCED, Maurice Strong, has chosen this location for her 'Sacred Earth Wisdom Keepers Circle,' where a fire will burn and a drum will be beaten without interruption, to provide a sacred dimension to the U.N. proceedings. She has invited friends, Native Americans, New Age prophets and astrologers, Gaia biologists and economists, and grass-roots activists.

"After the evening's fire ceremony Hanne explains her worldview to me. It is typical of New Age leaders I've met—a blend of Christian and Native American apocalyptic millennial prophecy, fortified by economic models of the 'Limits to Growth' kind, and pragmatic frontier spirit—every man for himself. In two years, she says, the American economy will be in shambles. Diseases that make AIDS look like the common cold will overrun the earth.

"Four and a half billion people will 'check out' over the next seven years. A few places will be safe to live, like Baca Valley, Colorado. She has pulled her guys together there—organic farmers, spiritual leaders. 'Come live with us', she offers."[418]

Clearly Hanne Strong's ancient spirit guides were wrong, as 4.5 billion people did not suddenly die between 1992 and 1999.

The Sacred Earth Gathering opened with Lucifer's Great Invocation (abridged here):

"Let light stream forth into human minds.

Let Light descend on Earth…

May the Coming One return to Earth…

From the centre where the Will of God is known

Let purpose guide all little human wills -

The purpose which the Masters know and serve…

"Let Light and Love and Power restore the Plan on Earth."

The report from Rio continues:

"Friday, June 5, 1992: Senator Al Gore opens the 'Forum for Spiritual and Parliamentary Change'. He speaks more passionately than when I last saw him, but can he articulate a green platform? Miles beyond the

---

418 Report from Rio: The Earth Summit, by Lavina Currier, Tricycle magazine, http://www.tricycle.com/feature/report-rio-earth-summit

others, but is he wily enough to take on the demon monkeys?"

The Forum for Spiritual and Parliamentary Change. Who knew? Nor is there any word on what the demon monkeys thought of Gore's speech.

The Strongs' Gathering issued a Declaration, published to all the delegates at Rio: "The responsibility of each human being today is to choose between the force of darkness and the force of light."

That depends, of course, on what you mean by "light".

"We must therefore transform our attitudes and values, and adopt a renewed respect for the superior law of Divine Nature."

In his opening address to the biggest collection of world leaders in history, Maurice Strong reminded them of that statement:[419]

"We are reminded by the Declaration of the Sacred Earth Gathering, which met here last weekend, that the changes in behaviour and direction called for here must be rooted in our deepest spiritual, moral and ethical values. We reinstate in our lives the ethic of love and respect for the Earth which traditional peoples have retained as central to their value systems. This must be accompanied by a revitalization of the values common to all of our principal religious and philosophical traditions."

The Earth Summit, said Strong "is not an end in itself, but a new beginning. The measures you agree on here will be but first steps on a new pathway to our common future… It is, therefore, of the highest importance that all Governments commit themselves to… implementation of Agenda 21.

"Our essential unity as peoples of the Earth must transcend the differences and difficulties which still divide us. You are called upon to rise to your historic responsibility as custodians of the planet in taking the decisions here that will unite rich and poor, North, South, East and West, in a new global partnership to ensure our common future…the dawning of a new world."

UN Secretary-General Boutros-Boutros Ghali also weighed in on a spiritual note, telling the delegates:

"My hope is that what I may call the 'spirit of Rio' – that is, the spirit of Planet Earth – will spread throughout the world. The spirit of Rio must embody the full awareness of the fragility of our planet. The spirit of Rio must lead us to think constantly of the future, our children's future."

Everywhere you looked, leaders and delegates were popping out of the

---

419 Opening address to Rio Summit, 3 June 1992, http://www.mauricestrong.net/index.php/opening-statement6

woodwork with occult or Freemasonic allusions. Portgual's president Mario Soares, for example, nailed his dreams to the mast of Jesuit mason Pierre Teilhard de Chardin – a darling of the Theosophy movement:

"I come here as the representative of a small European country which has a long history and which is proud of the contributions it has made to creating the civilization of the 'universal' of which Teilhard de Chardin spoke."

That civilisation of the universal was de Chardin's belief that Planet Earth was heading to an "Omega Point" where it would gain consciousness and talk to humans. He called it the 'noosphere' and said it would manifest as the 'Spirit of the Earth'.

"It can become explicit only when our consciousness has expanded beyond the broadening, but still far too restricted, circles of family, country and race, and has finally discovered that the only truly natural and real human Unity is the Spirit of Earth."[420]

In other words, just like the Lucifer disciples, Teilhard de Chardin claimed world consciousness had to be raised before the spiritual fireworks could begin.

"The sense of Earth is the irresistible pressure which will come at the right moment to unite them (humankind) in a common passion.

"We have reached a crossroads in human evolution where the only road which leads forward is towards a common passion ... To continue to place our hopes in a social order achieved by external violence would simply amount to our giving up all hope of carrying the *Spirit of the Earth* to its limits." [emphasis added]

When we get there, however, Teilhard says the future civilisation must, of necessity, be governed by "the world wide installation" of a regime which firstly "must be international and in the end totalitarian" and secondly, "it must be conceived on a very large scale".[421]

He saw this as inevitable once the "totalisation" – or what we would now call the globalisation – of humanity occurred. Some, he predicted, would try to resist, but in the end resistance would be futile – humanity would grow to "love" the Regime – what we might call the 'lie back and think of England' response:

"Men having at last understood that they are inseparably joined elements of a converging Whole, and having learnt in consequence to love

420 "Building the Earth", Pierre Teilhard de Chardin, p. 43
421 *Human Energy* by Pierre Teilhard de Chardin, 1937, pp133-134

the preordained forces that unite them, a natural union of affinity and sympathy [to the Regime] will supersede the forces of compulsion."[422]

These phrases being kicked around at Rio might have sounded like random, fluffy "mostly harmless" epithets from over-excited politicians and UN officials to the thousands of journalists covering the event, but those words were coded and their meanings – like everything else occult (which itself means 'hidden') – were not apparent to those not trained to hear them. Portugal's president did not just wake up that morning and dream up the name of a long-dead French freemason. The reference to the totalitarian universal civilisation was deliberate. Boutros-Boutros Ghali did not just randomly raise up the "Spirit of the Earth" out of nowhere either.

To journalists, it was a UN talkfest. To the public watching on TV screens around the world, it was both a spectacle and a talkfest, but also an awareness raiser. To the Luciferians, it was turning on a massive spiritual homing beacon, announcing to Satan/the Christ/Maitreya/Buddha/The God of All that his disciples had done their job, and the time was nearing.

"The 27 principles of the 'Rio Declaration', building on the Stockholm Declaration, clearly represent a major step forward in establishing the basic-principles that must govern the conduct of nations and peoples towards each other and the Earth to ensure a secure and sustainable future," said occultist and Rio Summit director Maurice Strong.

"I recommend that you approve them in their present form and that they serve as a basis for future negotiation of an 'Earth Charter', which could be approved on the occasion of the 50th anniversary of the United Nations."[423]

Strong also elaborated further on Agenda 21, the UN's "agenda" for the 21st century. Back in 1992 it was a twinkle in the occult team's eye. Now it's part of the law in your country.

"Agenda 21 is the product of an extensive process of preparation at the professional level and negotiation at the political level," Strong told delegates. "It establishes, for the first time, a framework for the systemic, co-operative action required to effect the transition to sustainable development. And its 115 programme areas define the concrete actions

---

422 "A Great Event Foreshadowed: The Planetisation of Mankind" in the *Future of Man* by P Teilhard de Chardin, pp 124-125. As cult-watcher and author of False Dawn, Lee Penn, points out, this was the end result in Orwell's *1984* as well. A version of Stockholm Syndrome.
423 http://www.mauricestrong.net/index.php/opening-statement6

required to carry out this transition. In respect of the issues that are still unresolved, I would urge you to ensure that the agreements reached at this historic Summit move us beyond the positions agreed by Governments in previous fora.

"The issue of new and additional financial resources to enable developing countries to implement Agenda 21 is crucial and pervasive. This, more than any other issue, will clearly test the degree of political will and commitment of all countries to the fundamental purposes and goals of this Earth Summit."

Delegates were told that for the sake of the planet they would have to commit to free trade as well.[424]

"The transition to a more energy-efficient economy that weans us off our overdependence on fossil fuels is imperative to the achievement of sustainable developments. The removal of trade barriers and discriminatory subsidies would enable developing countries to earn several times more than the amounts they now receive by way of Official Development Assistance. Large-scale reduction of their current debt burdens could provide most of the new and additional resources they require to make the transition to sustainable development through implementation of Agenda 21.

"We also need new ways of financing environment and development objectives. For example, emission permits that are tradeable internationally offer a means of making the most cost-effective use of funds devoted to pollution control while at the same time providing a non-budgetary means of effecting resource transfers. Taxes on polluting products or activities, like the CO_2 taxes now being levied or proposed by a number of countries, could also be devoted to financing of international environment and development measures. While none of these promising measures may be ripe for definitive action at this Conference, I would urge the Conference to put them on the priority agenda for the early post-Rio period."

Maurice Strong's next statement is the one you can beat him around the ears with.

"Over the next 20 years, more than one quarter of the Earth's remaining species may become extinct. And in the case of global warming, the

---

424 The Lucis Trust was also pushing free trade for the Satanist agenda: "The need for goodwill is evident also in the area of trade...hopefully they can eventually merge and can harmonise their arrangements so that goods and services can flow freely throughout the world." http://web.archive.org/web/19990225001028/http://www.lucistrust.org/goodwill/wgnl981.htm

Inter-governmental Panel on Climate Change has warned that if carbon dioxide emissions are not cut by 60 per cent immediately, the changes in the next 60 years may be so rapid that nature will be unable to adapt and man incapable of controlling them."

So how did his predictions stack up? Badly, as it turned out. Of the past 20 years, scientists have been unable to detect any global warming in the last 17 of them and peer reviewed climate scientists are now saying warming might not resume until nearly 2040. Carbon dioxide emissions were not "cut by 60 per cent immediately" but instead continued a straight-line increase upwards. Sea levels have risen around 3cm in 20 years, which extrapolates out to 15cm or six inches within a hundred years, nowhere near the 200cm the IPCC was predicting.

Of course, all these inconvenient 'truths' were far in the future. For now, in June 1992, Maurice Strong could bask in the afterglow of the Earth Summit, knowing he was going to be clipping the ticket for a very long time to come.

Satanist he essentially might have been, but Strong went on to become a senior advisor to the President of the World Bank, an advisor to Toyota Corporation, an advisor to Harvard University, the World Business Council for Sustainable Development, the World Wildlife Fund, the Eisenhower fellowships and many more.

Who are some of the other voices we hear from?

Another New Ager is Paul Ehrlich, cited here by fanatic David Suzuki in a Canadian article in 1990:

"Ehrlich concludes that it would be a dangerous miscalculation to look to technology for the answer to [environmental problems]. Scientific analysis points toward the need for a quasi-religious transformation of contemporary culture."[425]

If you've heard of "The Gaia Hypothesis", you'll probably know it as the theory put forward by scientist James Lovelock, that the earth is an interdependent organism with a life force of its own emanating from the eco-system.

Lovelock is right, and he's also wrong. The Earth is a lump of iron and rock in space that happens to have life on it and that life has adapted well to its environment. If a species becomes extinct, another takes its place in the ecosystem, and there are good physical explanations for the

---

425 "The Man Who Cries Wolf", by David Lees, *Harrowsmith* magazine, March/April 1990, pp34-44

various phenomena on the planet. The Earth itself doesn't care, only the life cares and only insomuch as it is self-aware.

Nonetheless the Gaia hypothesis is a good rough way of saying that we should look after our environment because negative impacts cause changes that we might not like down the track. Where it goes off the rails is during its hijacking as a religious concept.

What most of you probably don't know is that Lovelock is another New Age guru, a 'fellow' of the occult Lindisfarne Centre which has stated the whole planet will be forced to undergo the "Lucifer Initiation". It just goes to show what an impact on popular culture the Lucifer devotees have had.[426]

In an essay on "The Gaia Concept", journalist Cliff Kincaid explains how the science is twisted into the metaphysical:

"These people believe in Gaia – an 'Earth spirit,' goddess or planetary brain – and they think that human beings can have mystical experiences or a spiritual relationship with this entity. In order to protect Gaia, in their view, the U.S. and other industrial countries have to be prohibited from certain uses of the world's natural resources. This is called "sustainable development."[427]

"In general and secular terms, this cult, which combines aspects of the animal rights and radical environmentalist movements, holds that human beings are exploiting the Earth and other living creatures for selfish purposes."

Gaia the 'Spirit of the Earth', as the UN's Secretary-General Boutros-Boutros Ghali termed it in Rio, is only a façade however. As the creators of the New Age movement have said repeatedly in this book, all are one and the one is Satan:

"It is Satan who is the God of our planet and the only God."[428]

It might be unpalatable to casual New Age followers, who prefer Angel bling and feminine spirits like Gaia as the object of their allegiance, but sadly that's not how the theology works.

The way it has been sold is quite cunning, however. The idea of Gaia as a complete spiritual system is known as panentheism (as noted earlier

---

426 http://www.lindisfarne-association.org/links.html Other Lindisfarne fellows include Gnostic author Elaine Pagels
427 "Al Gore, The United Nations, and The Cult of Gaia" by Cliff Kincaid, 1999, http://www. usasurvival.org/cultofgaia.html
428 The Secret Doctrine by Helena Blavatsky, Vol. 2, Page 234
http://www.theosociety.org/pasadena/sd/sd2-1-13.htm

in this book), which literally means everything is in God. This stands in sharp discord with Christianity, where God exists outside Time and outside the Universe as a separate entity from creation.

Both versions of God cannot – by definition – be correct, which is why those who try and paint this movement as "Christian" in any way, shape or form are simply displaying their ignorance on spiritual teachings. Either God is part of the universe, or God is not. God cannot be both simultaneously, in just the same way that there is no such thing as a square circle.

Leave aside the issue of whether you personally have a religious belief of any kind, because it's irrelevant to the logic of the argument. To understand why the New Age and genuine grass-roots Christianity are on a collision course, you simply have to understand the differences.

Two stark choices. One group says the Coming One reveals himself to special elite people first, with trickle-down awareness on a need to know basis so as not to overwhelm "little human minds". The other group says Jesus Christ is God incarnate, not merely a "Master" or "Avatar" or 'enlightened one'. Jesus, they say, came to earth specifically to show the world precisely who he was and why people needed to believe exclusively in him: "I am the way, the Truth and the Life. No one comes to the Father but through me". This second group says the first group's offer of "all paths lead to God" is in direct conflict with the words of Jesus, who says only one path will get you there. They cannot both be right. The second group says Jesus came to save the world, not just special elite people like the Pharisees of his day.

That's the difference in a nutshell.

It is the philosophy of the first group, however, that now dominates popular culture and science – particularly climate and environmental science – a field with lots of 'special elite people' working in it. In March 2014 the UN climate body the IPCC will release its Fifth Assessment Report (AR5) and specifically the Working Group II studies. Chapter 13 of that document is entitled "Human Security", and the coordinating Lead Author of the team writing that chapter is Neil Adger.

Adger's blog for the World Bank lists his credentials,[429] including that he's a member of something called "The Resilience Alliance".

If you go to the Resilience Alliance website, you discover they endorse

---

something they describe as "Pan-archy". Panarchies draw, says the group, "on the image of the Greek god Pan – the universal god of nature. This hoofed, horned, hairy and horny deity (Hughes 1986) represents the all pervasive spiritual power of nature and has a personality and role that is described in sections of the Orphic Hymns as Goat-legged, enthusiastic, lover of ecstasy, dancing among stars, weaving the harmony of the cosmos into playful song (as translated by Hughes, 1986)."[430]

Let's remind ourselves what Pan looks like, shall we:

Looks a trifle familiar don't you think?

Pan, if you were not sure, is the source of the word "Panic", and was known as a god of rape, lust and deviance. He has been immortalised in stone by the ancients doing something unspeakable to a goat.

The Resilience Alliance philosophy is to prepare for moments of abrupt and sudden change to achieve revolution in society. Again, Pan to them symbolises the opportunity to strike the existing system when it is most vulnerable:

"Pan has a destabilizing role that is captured in the word panic, directly derived from one facet of his paradoxical personality. His attributes are described in ways that resonate with the attributes of the four phase adaptive cycle; as the creative and motive power of universal nature, the controller and arranger of the four elements – earth, water, air and fire (or perhaps, of K, alpha, r and omega!). He therefore represents the inherent features of the synthesis that has emerged in this comparison of ecological and social systems."

Good to know that climate science is in great hands.

You might recall from an earlier chapter seeing a Tarot card for the Magician illustrating the "As above, so below" spell. Above the Magician's head was an occult symbol of life, a number 8 on its side, the mathematical symbol for infinity.

The same diagram appears on the Panarchy website, as an obscure representation of their alpha to omega, revolt and remember strategy. Apparently it all makes sense if you're an initiate.

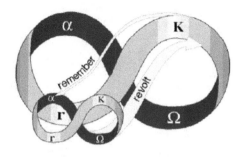

Panarchy seems to be the theme for some of the other New Age apostles out there as well.

Barbara Marx-Hubbard was once nominated as a Democrat Vice-Presidential candidate for the 1984 election. She has also served on the World Future Society with Maurice Strong and former US Defense Secretary and World Bank president Robert McNamara. She is also a prolific New Age author, who claims to have heard from a "spirit guide" that the Coming One intends to destroy vast swathes of humanity – half the world population – for the sake of the planet:[431]

"We are in charge of God's selection process for planet Earth," the spirit told Marx-Hubbard. "He selects, we [the Masters] destroy. We are the riders of the pale horse Death. We come to bring death to those who are unable to know God.

"The selection process must be made so that only the God-conscious receive the power of co-creators. We will use whatever means we must to make this act of destruction as quick and painless as possible to the one half of the world who are capable of evolving."

It's comforting to know that a woman who claims to speak to demons advocating the death of one half of humanity was a Democrat National Convention nominee for Vice-President. Again, you might think this is bat guano, but people with ideas like this are now in high places.

Marx-Hubbard is now 83, but she is so important to the New Age Community that guru Neale Donald Walsch has written a biography of her. The message she's preaching is that if the world converts to the New Age, the 'Masters' might spare them, the violence and destruction might be avoided:[432]

"In the eighties, although we had a conservative government, we began to have these great big peoples' events like the 1986 World Healing Meditation, 1987 was the Harmonic Convergence. Between 1988 and '90 Gorbachev led

---

431 The Book of Co-Creation by Barbara Marx-Hubbard, New Visions, 1980, p59
432 1993 interview with Thresholds magazine, http://www.som.org/3library/interviews/hubbard.htm

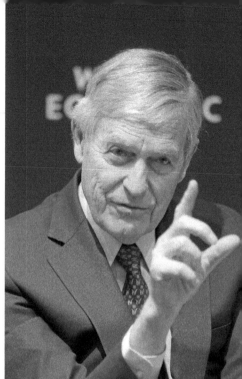

TOP LEFT: Business tycoon and New Ager Maurice Strong pictured at the 1972 Stockholm Environment Conference that set in motion the path to global governance on the back of environmental legislation. TOP RIGHT: Former Democrat Senator Tim Wirth now heads up Ted Turner's billion dollar United Nations Foundation. BOTTOM: New Ager Al Gore is another to allow his religious views to impact on his politics. That hasn't stopped him from making a fortune off the climate change scare.

the Communist empire into self-dissolving, the Berlin Wall came down, we had the second Earth Day. We had a real powerful breaking up of old forms. Now recently, the Arab-Israeli peace treaty. The consciousness is shifting. And I personally believe very deeply, not only in the personal change but in mass events, where the mass consciousness aligns itself and brings itself to bear on shifting the overall consciousness of the Earth. I think that that's happening. My own view is that we have the opportunity now, collectively, to shift the consciousness from fear to love by doing it ourselves en masse, and that we can avoid the violence if we do this."

US Vice President Al Gore was not really motivated by science when he produced *An Inconvenient Truth*. His book *Earth In The Balance* suggests his New Age religious views and hopes for a one world universal religion played a big part in his apocalyptic claims about climate change:

"But the emergence of a civilization in which knowledge moves freely and almost instantaneously through the world has ... spurred a renewed investigation of the wisdom distilled by all faiths. This panreligious perspective may prove especially important where our global civiilzation's responsibility for the earth is concerned."[433]

Gore, like the other leaders at the 1992 Earth Summit, channels freemason Teilhard de Chardin:

"The fate of mankind, as well as of religion, depends upon the emergence of a new faith in the future...Armed with such a faith, we might find it possible to resanctify the earth."

"We are close to a time when all of humankind will envision a global agenda that encompasses a kind of Global Marshall Plan, if you will, to address the causes of poverty and suffering and environmental destruction all over the earth," adds Gore.

Then there's Bill and Melinda Gates. The Microsoft billionaire turns out to be listed as a godfather on the Lucis Trust offshoot, "the New Group World Servers" website. After publicity in North America about the link, the page was removed from sight but is still available on the Internet Archive.[434]

Gates was also among a number of donors praised by the Lucis Trust for assisting in bringing back the "divine":

"Today there is a small but influential minority of responsible and wealthy citizens and organizations who are taking action and showing great leadership in alleviating the current sufferings. The ones we may

433 "Earth In The Balance" by Al Gore, Plume, 1993 pp258-259
434 http://web.archive.org/web/20100109052717/http://www.ngws.org/service/Financial-list.htm

have heard about are Bill and Melinda Gates Foundation, George Soros, John D. Rockefeller, Henry Ford Foundation, Andrew Carnegie Endowment, as well as many others. These philanthropic efforts are certainly making constructive and positive impacts as well as setting a necessary example to other wealthy billionaires on how money can be channelled towards constructive means to help rebuild a better civilization, restore the divine circulatory flow and heal and uplift people from the scourge of poverty, war and disease."[435]

The Lucis Trust is probably very happy with Bill Gates[436] for helping to roll out the Common Core education curriculum for American schools, which you will meet in the next chapter.

The Luciferians are also pushing the complete opening up of international borders to migrants. "As well as the free, unprejudiced flow of goods and services, the world would be a far better place if the free flow of individuals were possible.

"The world is currently very far away from this ideal... [which would bring] in its wake a great flourishing of cultural understanding and cross-fertilisation."[437]

Or, as some cynics might be saying, "more beheadings and home grown terrorism".

If you have waded through these pronouncements from the "enlightened ones", you might get some relief from a German phrase describing what is really happening as the Lucifer/Gaia/Pan movement deepens its efforts to literally seize control of the world:

"Geistige Umnachtung", which literally means "spiritual endarkenment", or a "spiritual nighttime".

If you are a *Lord of the Rings* fan, heed the words, *and in the darkness bind them.*

How far away are they from achieving this? Not far. They already have your public service, and your children...

---

435 http://www.lucistrust.org/fr/content/download/26079/330809/file/2010_Symposium_transcript.pdf
436 There's no suggestion (or at least I haven't seen any evidence of it) that Bill and Melinda Gates are followers of the Lucis Trust per se. They were being praised more in the vein of being 'useful idiots' for the wider cause
437 http://web.archive.org/web/19990225001028/http://www.lucistrust.org/goodwill/wgnl981.htm

CHAPTER 24

Controlling Children's Minds: The Modern Pied Pipers

"Nine he gave to Mortal Men, proud and great, and so
ensnared them. Long ago they fell under the dominion
of the One, and they became Ringwraiths, shadows
under his great Shadow, his most terrible servants. Long
ago. It is many a year since the Nine walked abroad. Yet
who knows? As the Shadow grows once more, they too
may walk again. But come! We will not speak of such
things even in the morning of the Shire..."

– J R R Tolkien, LOTR

It's one thing for your vehicle to have fuel in the tank. It's entirely another to make it ignite as required.

In UN terms, the long term planning that began with Alice Bailey's visions and the establishment of the Lucifer Publishing Company in 1922 took some time to come to fruition. It required adepts like Robert Muller and Dag Hammarskjöld to put programmes in place like UNESCO.

Simultaneously, the Marxist influence over education was decisive as well. Marx, Lenin, Trotsky and Gramsci had all emphasised the need to "break" middle class, "bourgeois" values and in doing so break the family. If you could break the family, you could control the future, the Marxists believed.

In his epic history of the 19th and 20th centuries, President Bill Clinton's former history professor at Georgetown University, Carroll Quigley, waxed lyrical at how the revolutionaries put the family and its values under sustained attack in the early 20th century. It's worth quoting extensively from one of America's top historians, because his is a voice from 1966 and it bears eerie parallels with the predicament we are now in.[438]

"This disintegration of the middle classes had a variety of causes... many of them going into the very depths of social existence. All these causes acted to destroy the middle classes by acting to destroy the middle class outlook. And this outlook was destroyed, not by adult middle class persons abandoning it, but by a failure or inability of parents to pass it on to their children.

"This failure was thus a failure of education, and may seem, at first glance, to be all the more surprising, since our education system has been, consciously or unconsciously, organised as a mechanism for indoctrination of the young in middle class ideology."

Remember, Professor Quigley was writing this in the mid 1960s. He was not privy to some of the deliberate subversions of education curricula that have now become apparent. He certainly does not appear to have been fully aware of the agenda of the Lucis Trust, or its influence via the UN on world education and social policy, although as we have previously seen he was aware that a "secret" elite was attempting to change the world, with its roots in freemasonry.

"The chief external factor in the destruction of the middle class outlook has been the relentless attack upon it in literature and drama through most of the twentieth century," wrote Quigley.

"In fact, it is difficult to find works that defended this outlook or even assumed it to be true, as was frequent in the nineteenth century. Even those writers who explicitly accepted the middle class ideology...tended to portray middle class life as a horror of false values, hypocrisy, meaningless effort and insecurity.

"In the earlier period, even down to 1940, literature's attack on the middle class outlook was direct and brutal, from such works as Upton Sinclair's *The Jungle* or Frank Norris' *The Pit*, both dealing with the total corruption of personal integrity...

---

438 The extensive passages that follow, except where indicated otherwise, are taken from *Tragedy And Hope* by Carroll Quigley, Macmillan NY 1966, pages 1247-1270. You can buy the book on Amazon or read it here: http://www.carrollquigley.net/pdf/Tragedy_and_Hope.pdf

"These early assaults were aimed at the commercialisation of life under bourgeois influence and were fundamentally reformist in outlook because they assumed that the evils of the system could be removed, perhaps by state intervention.

"By the 1920s the attack was much more total, and saw the problem in moral terms so fundamental that no remedial action was possible. Only complete rejection of middle class values could remove the corruption of human life seen by Sinclair Lewis in *Babbitt* or *Main Street*.

"After 1940, writers tended less and less to attack the bourgeois way of life; that job had been done. Instead they described situations, characters and actions that were simply nonbourgeois: violence, social irresponsibility, sexual laxity and perversion…human weakness in relation to alcohol, narcotics or sex, or domestic and business relationships conducted along completely nonbourgeois lines."

Quigley argues that some writers like Ernest Hemmingway abandoned any pretense of holding to middle class values, "embracing the cult of violence" in his writing and his life. Quigley was thus not surprised that "when his virility, in the crudest sense, was gone, he blew out his brains."

In essence, Quigley's argument is that the literati deliberately destroyed family values by presenting them as hopeless, unattainable and hypocritical. Far easier, said twentieth century writers, to go with the flow. The future was presented as a hopeless place, and the dreams of middle class youth were presented as unfulfilled.

"It often took the form, in more recent times, of a rejection of a man's whole life achievement by his sons, his wife or himself."

Think *Death of a Salesman*, the Pulitzer and Tony-award winning play by Arthur Miller, taught in high schools around the world, and you have the epitome of the genre.

Students, Quigley says, were taught to give in to their instant desires rather than tough anything out:

"Novel after novel, or play after play, portrayed the wickedness of the suppression of good, healthy, natural impulse and the salutary consequences of self-indulgence, especially in sex. Adultery and other manifestations of undisciplined sexuality were described in increasingly clinical detail and were generally associated with excessive drinking or other evasions of personal responsibility, as in Hemingway's *A Farewell To Arms* and *The Sun Also Rises*, or in John Steinbeck's love affair with personal irresponsibility in *Cannery Row* or *Tortilla Flat*.

"The total rejection of middle class values, including time, self discipline and material achievement, in favour of a cult of personal violence was to be found in a multitude of literary works from James M Cain and Raymond Chandler to the more recent antics of James Bond. The result has been a total reversal of middle class values by presenting as interesting or admirable simple negation of these values by aimless, shiftless and totally irresponsible people.

"A similar reversal of values has flooded the market with novels filled with pointless clinical descriptions, presented in obscene language and in fictional form, of swamps of perversions ranging from homosexuality, incest, sadism, and masochism, to cannibalism, necrophilia and coprophagia.

"These performances, as the critic Edmund Fuller has said, represent not so much a loss of values as a loss of any conception of the nature of man…From this has emerged the Puritan view of man (but without the Puritan view of God) as a creature of total depravity in a deterministic universe without hope of any redemption."

But it is not just literature and the arts which Professor Quigley argued had poisoned the baby boomer generation and its successors, so too were new "child centred" educational theories of the type already alluded to by the UN's Robert Muller and the Lucis Trust, with their emphasis on the child finding themselves, unlocking their mysteries as to who they are in the grand design of the universe.

"By 1910 or so, childrearing and educational theories had accepted the idea that man was a biological organism, like any animal, that his personality was a consequence of hereditary traits and that each child had within him a rigid assortment of inherited talents and a natural rate of maturation in the development of these talents. These ideas were incorporated in a series of slogans of which two were: 'Every child is different,' and 'He'll do it when he's ready'.

"From all of this came a wholesale ending of discipline, both in the home and in school, and the advent of 'permissive education' with all that it entailed. Children were encouraged to have opinions and speak out on matters of which they were totally ignorant; acquisition of information and intellectual training were shoved into the background.

"All this greatly weakened the disciplinary influence of the educational process, leaving the new generation much less disciplined, less organised and less aware of time than their parents."

The new educational theories, said Quigley, favoured girls over boys and resulted in a major blow to the self-esteem of boys which, he argued, had a serious impact on society overall.

"By the age of ten or twelve, girls were developed physically, neurologically, emotionally and socially about two years in advance of boys. All this tended to make boys less self-assured, indecisive, weak, and dependent. The steady increase of women teachers in the lower grades worked in the same direction, since women teachers favoured girls and praised those attitudes and techniques that were more natural to girls.

"New methods, such as the whole-word method of teaching reading or the use of true-and-false or multiple-choice examinations, were also better adapted to female than to masculine talents. Less and less emphasis was placed on critical judgment, while more and more was placed on intuitive or subjective decisions."

That last point, the deliberate skewing of education towards the subjective, will be covered off shortly. We now know why that was done, and you will see examples of it in action.

"In this environment," wrote the Professor, "girls did better, and boys felt inferior or decided that school was a place for girls and not for boys. The growing aggressiveness of girls pushed these hesitant boys aside and intensified the problem. As consequences of this, boys had twice as many 'non readers' as girls, several times as many stutterers, and many times as many teenage bedwetters."

This gender insecurity has translated in relationships, Quigley wrote, where men felt unsure of their masculinity and women resented them for their weakness, compounding the problem. The family unit was under siege in 1966 like never before. If you fast forward nearly fifty years, what would Quigley have made of society's continued slide into the abyss since then? One can only speculate.

In perhaps the most ironic statement of all, coming from one of America's most celebrated university professors paid to teach people into their 20s, Carroll Quigley noted that the push to keep children in higher education for much longer – school ended around age 13 in the 1880s compared with a university degree in the sixties – was also hostile to the family unit:

"The lengthening of the interval of time between sexual awareness and the ending of education, from about two years in the 1880s to at least ten or twelve years in the 1960s, has set up such tensions and strains in the bourgeois American family that they threaten to destroy the family and are already in the process of destroying much of the middle class outlook...from this has emerged an almost total breakdown of communication between teenagers and their parents' generation.

"When any effort is made too talk across the gap between the generations, words may pass but communication does not. Behind this protective barrier a new teenage culture has grown up. Its chief characteristic is rejection of parental values and of middle class culture.

"In many ways this new culture is like that of...tribes: its tastes in music and the dance, its emphasis on sex play, its increasingly scanty clothing, its emphasis on group solidarity, the high value it puts on interpersonal relations (especially talking and social drinking), its almost total rejection of future preference and its constant efforts to free itself from the tyranny of time.

"There is widespread tolerance and endless discussion of all these issues. This discussion, like most of the adolescents' endless talk, never reaches any decisions but leaves the question open or decides that 'it all depends how you look at it'.

"As part of such discussions, there is complete casual frankness as to who has had or is having sexual experiences with whom. Widely permeated with an existentialist outlook, the adolescent society regards each sexual experience as an isolated, contextless act, with no necessary cause or consequence, except the momentary merging of two lonelinesses in an act of togetherness.

"Among middle class youth it is accompanied by an atmosphere of compassion or pity rather than of passion or even love...Among lower class persons it is much more likely to be physiologically inspired and associated with passion or roughness. This often attracts middle class girls who become dissatisfied with the weakness and undersexuality of middle class boys."

From the fifties and sixties onward, Quigley writes, youth have abandoned the basic tenets of western civilisation in favour of a doctrine of tolerating virtually anything, primarily to feel "inclusive" or part of a group. We have become addicted to "belonging" he says, without really caring what it is we actually belong to.

"These urges [to help others or take part in popular causes] are existentialist. They give rise to isolated acts that have no significant context. Thus an act of loving or helping has no sequence of causes leading up to it or of consequences flowing from it. It stands alone as an isolated experience of togetherness and of brief human sharing.

"Today's youth has no concern for the whole picture; they have rejected the past and have very little faith in the future. Their rejection of intellect and their lack of faith in human reason gives them no hope that any meaning can be found for any experience, so each experience becomes an end in itself, isolated from every other experience."

In other words, in the latter half of the twentieth century, we had created a major civilisation with a spiritual vacuum – people with no real meaning in their lives. If you recall the history lesson surrounding the first migrants to America and how tough they had to be, you can see how in the space of just 300 years those cultural attributes of brutal toughness had largely disappeared.

Little wonder that the UNESCO team employed by Robert Muller and dispersing their education theology across the world, were able to step into the void and use it to their advantage. While many were aware of creeping inroads being made by the new ideologies, they culminated in that planetary shift of consciousness many of us witnessed in Rio. The 1992 Earth Summit in Brazil was one giant leap forward into filling that void, with the ratification of the global masterplan, Agenda 21.

Within Agenda 21 itself, the policy calls for re-education of the public, from schoolchildren to pensioners, to turn them into people who will accept the new world order, and uphold its ideals:

"Education, including formal education, public awareness and training should be recognized as a process by which human beings and societies can reach their fullest potential. Education is critical for promoting sustainable development and improving the capacity of the people to address environment and development issues.

"While basic education provides the underpinning for any environmental and development education, the latter needs to be incorporated as an essential part of learning. *Both formal and non-formal education are indispensable to changing people's attitudes* so that they have the capacity to assess and address their sustainable development concerns. It is also critical for achieving environmental and ethical awareness, values and attitudes, skills and behaviour consistent with sustainable development and for effective public participation in decision-making.

"To be effective, environment and development education should deal with the dynamics of both the physical/biological and socio-economic environment and human (which may include spiritual) development, should be integrated in all disciplines, and should employ formal and non-formal methods and effective means of communication."[439]

As a key objective, Agenda 21 lists:

"To achieve environmental and development awareness in all sectors of society on a world-wide scale as soon as possible."[440]

"To strive to achieve the accessibility of environmental and development education, linked to social education, from primary school age through adulthood to all groups of people;

"To promote integration of environment and development concepts, including demography, in all educational programmes, in particular the analysis of the causes of major environment and development issues in a local context."

In other words, indoctrination to lay the groundwork and justification for the Regime that is to come.

The Agenda 21 document calls for primary school children, in particular, to be targeted, but right through to tertiary level and beyond. It calls for the introduction of charter schools, or what it calls "small, grass roots" institutions, and a "lifting of restrictions on private schooling" to make this possible by way of increased taxpayer funding of NGOs.

One of the first tentacles of the Agenda 21 brainwashing came from the International Baccalaureate education system, which provides programmes from primary through secondary education. Nearly 1,500 American schools have signed up to the IB[441], 167 UK[442], 151 Australian schools[443], 328 Canadian[444] and 21 New Zealand schools.[445] The baccalaureate system is UN endorsed and heavily based on creating "global citizens".

"The IB is more than its educational programmes and certificates. At our heart we are motivated by a mission to create a better world through education…We promote intercultural understanding and respect, not as an alternative to a sense of cultural and national identity, but as an essential part of life in the 21st century.

"Our four programmes for students aged 3 to 19 help develop the intellectual, personal, emotional and social skills to live, learn and work in a

---

439 http://sustainabledevelopment.un.org/content/documents/Agenda21.pdf paragraph 36.3
440 Ibid, para 36.4
441 http://www.ibo.org/country/US/index.cfm
442 http://www.ibo.org/country/GB/index.cfm
443 http://www.ibo.org/country/AU/index.cfm
444 http://www.ibo.org/country/CA/index.cfm
445 See http://www.ibo.org/country/NZ/index.cfm

rapidly globalizing world. Founded in 1968, we currently work with 3,663 schools in 146 countries to develop and offer four challenging programmes to over 1,132,000 students aged 3 to 19 years."[446]

Here's an example of the introductory documents a teacher is required to know before teaching five year olds in a Baccalaureate school:[447]

SUSTAINABILITY AS INTERNATIONAL MINDEDNESS

Participants are requested to bring (either download to laptop or bring a hard copy):

- A copy of your school's programme of inquiry (POI)
- A few examples of units of inquiry
- Any documents your school has developed related to sustainability
- Any images or DVDs of your school related to sustainability that you would like to share
- Agenda 21 http://www.unep.org/Documents.Multilingual/Default.Print. asp?documentid=52 (See Children and Youth; Promoting Education)
- Decade for sustainability http://www.unesco.org/en/esd/
- Earth Charter http://www.earthcharterinaction.org/content/pages/ The-Earth-Charter.html
- UNESCO education statement http://www.unesco.org/en/esd/pro-gramme/educational-dimensions
- Visualising sustainability http://computingforsustainability.wordpress. com/2009/03/15/visualising-sustainability/
- Millennium assessment http://www.maweb.org/en/index.aspx
- Living Beyond Our Means: Natural Assets and Human Wellbeing http://www.maweb.org/en/BoardStatement.aspx
- Ecosystems and Human Wellbeing http://www.maweb.org/docu-ments/document.356.aspx.pdf
- Global status of ecosystem services http://www.wri.org/publication/ content/7755
- Biogeochemical Cycles of Nature (general reference) http://www. smithlifescience.com/CyclesNature.htm
- Developing a transdisciplinary programme of inquiry
- The Primary Years Programme as a model for transdisciplinary learning
- Making the PYP Happen: a curriculum framework for international primary education

446 International Baccalaureate website, http://www.ibo.org/general/who.cfm
447 Workgroup introduction for Primary Years Programme (PYP) teachers, http://www.ibo.org/iba/pyp3.cfm

- The Learner Profile Booklet
- A continuum of International learning
- Science and Social studies scope and sequence
- Other documents that you think may be relevant to you programme (Basis of Practice, CAS documentation)

First they came for the teachers, then they came for your children. The irony is that parents pay tens of thousands of dollars to have little Amanda or Timothy educated in a full IB school. You have to give the Devil his dues, it seems.

In New Zealand however, whilst many of our so-called top schools are offering the Baccalaureate curriculum, the State schools with their national standards NCEA (national certificate of educational achievement) are just as bad. You can see just how indoctrinated Generations X and Y have become, from an "exemplar" standard NCEA Level 2 paper (Year 12), passed with excellence by the student involved.

The paper was called "Level 2 Education for Sustainability 2008", and asked students to "describe worldviews, their expression through practices and activities and consequences for a sustainable future."

Students are informed in the exam paper that "Over time, people have developed many world views that have influenced the way people relate to and understand the environment. These world views have had consequences for a sustainable future by influencing people's beliefs, values, attitudes, practices and activities."

Background

Over time, people have developed many world views that have influenced the way people relate to and understand the environment. These world views have had consequences for a sustainable future by influencing people's beliefs, values, attitudes, practices, and activities.

Below are examples of **world views** that have influenced attitudes toward the natural world.

Religions	Indigenous traditions
• Judaeo-Christian	• Māori
• Islam	• Native American
• Hinduism	• Aboriginal Dreamtime
• Buddhism	
• Jainism	

Scientific world views	Philosophies
• Western Scientific	• Marxism
• The Gaia Hypothesis	• Capitalism
	• Deep Ecology
	• The Land Ethic

A **sustainable future** requires the development of ways of thinking and acting to meet the needs of the present generation without compromising the ability of future generations to meet their own needs.

Examples are then given of different world views for the student to consider:

The student is then instructed by the exam paper that "A sustainable future requires the development of ways of thinking and acting to meet the needs of the present generation without compromising the ability of future generations to meet their own needs."

It is a bald statement of fact, with no room for debate, and the definition is almost word for word based on Maurice Strong and Gro Harlem Brundtland's definition of sustainability in the 1987 Brundtland Commission report, later adopted at the Rio Earth Summit. There's a perfect example of someone "thinking globally" and it turns up at your local school.

Students are told that the four aspects of sustainability are "environmental or ecological, social, cultural and economic". When considering whether something is sustainable, the exam paper then instructs students to keep in mind "these useful concepts":

- Biodiversity
- Personal and social responsibility
- Interdependence
- Future generations
- Values and beliefs
- Kaitiakitanga [an NZ Maori word meaning guardianship, and a good example of Agenda 21 being locally-adapted]
- Equity

The student who sat this exam chose to contrast the "Indigenous" worldview with the "Western scientific" worldview. Of the "Indigenous", he/she said:

"This worldview is centred around the belief that humans are the 'guardians' of Earth; its land, animals, plants, environment. It is a spiritual worldview, believing in polytheism – in other words, the worshipping of various and many deities that have dominion over various and many areas of life.

"All of Earth's resources are viewed as gifts; taongas [Maori word for treasures that originally actually meant 'property procured by the spear'[448] – probably not what the student thought it meant], therefore to

---

448 "A Dictionary of the New Zealand Language" by William Williams, 1852 edition. See also

indigenous cultures/ethnic groups, they must be respected and valued. "The relationship between Earth and humans is of a reciprocal kind; basically, 'I'll take care of you if you take care of me'. This promotes inter-connectedness and the belief that man and Earth are a 'whole'."

Now listen to the language the student uses to describe the "Western scientific" worldview, his or her own culture:

"Western scientific (with Judeo-Christian values/capitalist values). This worldview is centred around the belief that humans have dominion over Earth and its inhabitants, resources. They have the right to dominate all walks/forms of life.

"It is a monotheic worldview; this being founded on Judeo-Christian values in which one supreme being (ie God) gave the earth to humans to use.

"All of Earth's resources are seen as objects to exploit, manipulate and control, therefore having no other importance than to meet man's needs and wants, becoming subservient.

"The relationship between Earth and humans is of a one way kind; basically, 'I'll exploit you, if you can provide for me'. This promotes a compartmentalised approach in that everything is categorised, having no linkage to each other and therefore having no effect on each other."

The examiner of the paper remarks in the margin, "The candidate has an understanding of worldviews that goes beyond the descriptive. They understand the connection between western scientific and Judeo-Christian beliefs."

Little wonder students are coming out of high schools with cultural self-loathing, with this level of educational input.

Over the page, in the next question, the student paints an idealistic Agenda 21-derived image of indigenous culture in action:

"Indigenous practices/activities are focused on 'guarding' and 'maintaining' the land and environment. This can be seen in such practices as replanting trees, shrubs and vegetation, self-sufficiency (growing own food, collecting own water) and bartering. Replanting vegetation, in relation to sustainability allows a continuation and preservation of biodiversity and resources, in this case wood."

Tell that to Australia's aboriginal people and the New Zealand Maori, responsible for large deforestation and rapid species extinction. Reportedly

---

"Vocabulary of New Zealand" by Thomas Kendall, 1820

90,000 moa (what was the largest bird in the world) were slaughtered and eaten at the mouth of just one New Zealand river by indigenous tribes.[449] Tasty? Maybe. Sustainable? Not on your life. In many tribal villages around the world before Western civilisation freed them, women and children were frequently raped at will and prisoners and slaves were cannibalised. Once an area had been stripped of resources, the tribe moved on to another.[450]

Not only is the student in this exam ignorant about the realities of indigenous villages, he or she is also deeply ignorant of their own nation's past. All courtesy of an Agenda 21-skewed education system.

American thinker William Buckley Jr once wrote about how you could prove when indoctrination was successful:

"In the hands of a skillful indoctrinator, the average student not only thinks what the indoctrinator wants him to think . . . but is altogether positive that he has arrived at his position by independent intellectual exertion. This man is outraged by the suggestion that he is the flesh-and-blood tribute to the success of his indoctrinators."[451]

If that's you, ponder the evidence as we continue.

Question three of this New Zealand State-controlled exam shows precisely what children are being taught in state schools – quotes from discredited "population explosion" author Paul Erhlich, specifically this statement:[452]

"Our species is overshooting the capacity of its planetary home to support it in the long run...we have utterly changed our world; now we'll have to see if we can change our ways."

The student is asked to discuss the ramifications of that statement against each of the worldviews he/she had previously described. The student, of course, has been led by the nose to the only conclusion he/she could reach based on the biased data given:

"Obviously the indigenous world view is the most beneficial for a sustainable future. Not only does it positively apply to the environmental, social, economic and cultural aspects of sustainability, but it gives us the opportunity to change our ways of how we live.

---

449 "Penguin History of New Zealand" by Michael King, p63
450 See "The Great Divide: The Story Of New Zealand & Its Treaty" by Ian Wishart, Howling At The Moon, 2012
451 "Up From Liberalism" by William F. Buckley Jr., 1959
452 Cited as the statement to discuss on the exam paper, and drawn from "One With Nineveh" by Paul and Anne Ehrlich, Island Press, 2005, p 287

"In terms of environmental consequences, the indigenous worldview allows us to directly control and monitor our waste and pollution, leading to a decline in contamination, greenhouse gases and possibly illnesses and diseases. The future generations will be able to continue this effect and live in a less wasted and polluted world."

Again, utterly wrong. If indigenous tribes were truly living healthier lives, why was their infant mortality and general mortality shockingly high, and why are all the UN-affiliated charities constantly urging us to donate a dollar a day to provide drinking water, sewage and health systems for indigenous tribes?

The student then expresses a clear interest in Marxist wealth re-distribution:

"A social consequence is fair distribution/sharing of resources and opportunities. This means that no one will be better off than any other person; there will be equality between people and their share of resources and opportunities."

Of the western scientific worldview, the student says bluntly:

"It continues to threaten our world and our ways of life...[and has] arguably produced some of the 21st century's biggest environmental concerns, climate change, global warming, contamination of resources.

"Big oil, petroleum companies are 'exhaling' toxic fumes and chemicals into the sky and into our valuable atmosphere without a care in the world, causing excess amounts of CO_2 and therefore excess amounts of greenhouse gases. This attitude is what's going to be the driving force behind our demise. We are 'hurting' the planet that we inhabit, we are 'interconnected, linked.

"In conclusion, the consequences of the indigenous worldview, in my opinion, outweigh those of the western scientific worldview, in regards to a sustainable future for our subsequent generations."

The student got an "Excellence" for parroting the Agenda 21 material. So did another student who compared capitalism with indigenous. Of capitalism, she/he wrote:

"Its view is that the economy, society and the environment are in no way connected, and therefore not affecting one another."

Obviously it's a caricature of capitalism she/he has been taught, but look what else creeps in – a clear understanding of indigenous spiritualism. The UN's Robert Muller spoke all those decades ago in New York, and now here are his words being repeated by a New Zealand student 12,000 km away:

"The indigenous worldview (aboriginal, native American) is a very spiritual, harmonious worldview. The people are polytheistic, believe all things are interconnected spiritually and all are affected by each other.

"The environment, society and economy are linked and each individually important and highly significant to the other. It believes without a sustainable environment to live in, work in and exist in, both society and the economy will be greatly endangered. If each are treated with respect then no problems will occur and there will be no need for monetary solutions."

The student then embarks on an excellence-attaining essay based on Paul Ehrlich's claims, where she/he rubbishes capitalism and argues the world must embrace indigenous living styles in the name of equity with the third world.

You can read these papers in the link below.[453]

Again, one can only say "the fox is in the hen house". Those students are voters now. Their views were so similar they are clear evidence of the "group think" that has been deliberately unleashed in world education systems. The work of the Lucis Trust to prepare the world's youth to psychologically accept "the Coming One" is almost complete.

But not quite.

"We have laid a firm foundation for the future," admits UN Secretary-General Ban Ki-moon. "When we began, climate change was an invisible issue. Today, we have placed it squarely on the global agenda. When we began to work together, nuclear disarmament was frozen in time. Today, we see progress. We have advanced on global health, sustainable development and education."[454]

One of the stumbling blocks was to get Americans to accept UN ideals. It needed to be done through education, but the education systems were controlled by the individual state governments. The election of President Barack Obama and his "hope and change" team altered the balance of power. Now, just as they have been rolled out elsewhere in the world, aspects of Luciferian Robert Muller's education World Core Curriculum indoctrination have been incorporated into the US federal government's new "Common Core" education curriculum (similar to the national standards NCEA programme in New Zealand you've just become acquainted with).

453http://www.investigatemagazine.co.nz/Investigate/4636/common-core-indoctrination-of-children-surfaces-in-nz/
454 http://www.newstimeafrica.com/archives/20794

Common Core training documents from UNESCO have been supplied to ministries of education around the world. Countries have been told that as their governments have signed up to the UN's Millennium Development Goals, they need to incorporate common core into their education systems.

The training documents offer education officials a "Master's degree" qualification if they complete the course satisfactorily, which they can then take back and implement the policies in their home regions. The reasons for Common Core are simple, says UNESCO: global migration means different cultures are now in constant contact and education must be standardised throughout the world to acknowledge that.

"Demographic changes, migration, globalization and technological changes" are all relevant, says Common Core. "The fourth part of the course presents the main development frameworks, including the 'new' international commitments as part of Education for All/ Fast Track Initiative, Poverty Reduction, Millennium Development Goals, and discusses their impact on the role and methods/instruments of educational planning."[455]

As part of Common Core's introduction in the US, the federal government is empowering schools to collect and store biometric data on children in what is effectively a national database. That's fingerprints and retina scans. Every child goes to school, every child gets fingerprinted and identified for life, without having committed any crime.

After reading about the philosophy behind the oft-quoted "New World Order", do you still believe that if you have nothing to hide you have nothing to fear? Maybe you just didn't realise what it was that someone in the future might be looking for.

The people at the United Nations who dreamt all this up seriously believe Satan is returning in glory to govern the world. Do you want to give those people any surveillance hold over your family whatsoever? Even if they only manage to wheel out a $9.99 inflatable Satan from a dollar shop somewhere, it's their intentions that are dangerous, not how big their mascot is. Let's face it, if their mascot is real then that's a whole new ballgame altogether.

The infrastructure they are putting in place behind their new spiritual leader, real or not, is the issue we are dealing with today.

---

455 http://cherilyneagar.com/wp-content/uploads/fileuploads/2013/04/Course_outline_2011-1.pdf

They won't be looking for actual crimes, necessarily, but "thought crimes": resistance to the new regime, necessitating a "corrective treatment" programme at a facility somewhere.

"Keep an eye on this kid, he's not toeing the school line on indigenous cultural values. Must be an issue at home, let's talk to the parents."

"The standards themselves are written in a manner that will teach students what to think, not how to think," explains Mary Black, a teacher with four decades in the profession under her belt. "Most of the country hasn't yet felt the iron grasp of Common Core, but it's real and it is a threat."[456]

Dr Duke Pesta, an English professor at the University of Wisconsin, told *New American* magazine he's gravely concerned at the wholesale takeover of US education by Common Core:

"Yet another threat posed by Common Core is the absolute appropriation of Soviet ideology and propaganda in the constructing of Common Core and its implementation...A year from now, a year and a half from now, it will be so entrenched in the schools that there can be no modifications."[457]

An American history teacher penned an open letter to the people of the United States, warning about the introduction of Common Core. It's worth reading in full:[458]

A History Teacher's Message to America About Common Core Standards
by C.E. White

This week, President Obama will be sworn into office as the 45th President of the United States of America.

As a history teacher, I was elated to learn he would be placing his hand on two Bibles, one belonging to Dr. Martin Luther King, Jr., and the other belonging to President Abraham Lincoln, when he takes the oath of office to lead our great nation. Dr. King and President Lincoln helped define civil rights for America...historical heroes who transformed the idea of justice and equality.

As jubilant as I am that President Obama is symbolically using the bibles

456 "Educators Expose Dangers of Common Core National Education," by Alex Newman, New American, 24 September 2013, http://www.thenewamerican.com/culture/education/item/16610-educators-expose-dangers-of-common-core-national-education
457 Ibid
458 http://whatiscommoncore.wordpress.com/2013/01/20/history-teacher-speaks-out-stop-common-core/

of two of the greatest Americans in our nation's history, I am saddened that this administration seems to have forgotten what Dr. King and President Lincoln promoted regarding education.

In Dr. King's "Letter from the Birmingham Jail," he stated "the goal of America is freedom." As a teacher, it is such an honor to teach America's children about freedom and patriotism. However, over the past few years, I began to learn about a new education reform initiative called Common Core Standards. A few years ago, when I first heard of Common Core, I began doing my own research. My students represent the future of the United States of America, and what they learn is of utmost importance to me. I care about their future, and the future of our country.

My research of Common Core Standards kept me awake at night, because what I discovered was so shocking. I discovered that Common Core Standards is about so much more than educational standards. I wanted so badly to believe these changes would be good for our children. How can "common" standards be a bad thing? After all, isn't it nice to have students learning the same exceptional standards from Alabama to Alaska, from Minnesota to Massachusetts?

As a teacher, I began to spend nights, weekends, summers, even Christmas Day researching Common Core, because these reforms were so massive and were happening so quickly, it was hard to keep up with how American education was being transformed. I quickly began to realize that the American education system under Common Core goes against everything great Americans like Dr. King and President Lincoln ever taught. The very freedoms we celebrate and hold dear are in question when I think of what Common Core means for the United States.

One of my favorite writings about education from Dr. King is a paper entitled "The Purpose of Education." In it, he wrote "To save man from the morass of propaganda, in my opinion, is one of the chief aims of education. Education must enable one to sift and weigh evidence, to discern the true from the false, the real from the unreal, and the facts from the fiction."

When I sit in faculty meetings about Common Core, I hear "curriculum specialists" tell me that Common Core is here to stay and I must "embrace change." I am forced to drink the kool-aid. These specialists don't tell us to search for facts about Common Core on our own, they simply tell us what the people paid to promote Common Core want us to know. Didn't Dr. King want us to separate facts from fiction? Why are we only given information from sources paid to say Common Core is a good thing? Isn't that the exact same type of propaganda Dr. King discussed in his writings about education? Shouldn't we discuss why thousands of Americans are calling for a repeal of the standards?

I am told that I must embrace Common Core and I infer that resisting the changes associated with Common Core will label me "resistant to change." As a teacher, I definitely believe our classrooms are changing with the times and I am not afraid of change. Teachers across America are hearing similar stories about how they should "feel" about Common Core. This is a brainwashing bully tactic. It reminds me of my 8th graders' lesson on bullying, when I teach them to have an opinion of their own. Just because "everyone's doing it," doesn't make it right. In regards to Common Core, I am not afraid of change. I am just not going to sell-out my students' education so that Pearson, the Gates Foundation, David Coleman, Sir Michael Barber, Marc Tucker and others can experiment on our children.

I agree with Dr. King, which is why I am so saddened at how propaganda from an elite few is literally changing the face of America's future with nothing more than a grand experiment called Common Core Standards. Our children deserve more. Our teachers deserve more. Our country deserves more. Education reform is the civil rights issue of our generation, and sadly, parents, teachers, and students have been left out of the process.

President Lincoln once said "the philosophy of the classroom today, will be the philosophy of government tomorrow." With Common Core, new standardized tests have inundated classrooms with problems of their own. Teachers find themselves "teaching to the test" more and more. These tests violate our states' rights. I wonder if parents realized that all states aren't created equal in Common Core tests? Shouldn't all states, under "common" standards for everyone have everyone's equal input on how students are tested?

What about privacy under Common Core? Why didn't local boards of education tell parents about the changes to the Family Educational Rights and Privacy Act? Do parents realize their child's data, including biometric data such as fingerprints and retinal scans, is being placed in a state longitudinal data system and shared with others?

If our philosophy of the classroom is to violate states' rights, use children and teachers as guinea pigs, and hide from parents the fact that their child's data is no longer private, it can only be inferred that the philosophy of government tomorrow will do the same. What is America becoming?

As I watched President Obama place his hand on the bibles of Dr. King

and President Lincoln, the history teacher in me was overjoyed to watch such a patriotic moment in U.S. history. And yet, I was crushed at the realization that if we do not stop Common Core and preserve the United States educational system, the philosophy of our government tomorrow will not be the America we know and love.
    C.E. White

In Utah, mother Christel Swasey received confirmation from state educational authorities that Utah was implementing compulsory tracking of all students through its databases, and that Utah had changed the law, removing the right of parents to object to their children being tracked. The information is being added into the "longitudinal data system" and will follow preschoolers all the way to the workforce and beyond. Parental and child consent is not required if teachers, or anyone the database is shared with in a public/private "partnership" wants to access "personally identifiable information (PII)" on a child.[459]

You saw examples in the New Zealand NCEA papers quoted earlier of lessons becoming exercises in "group think" based on the national curriculum. American teacher Diana McKay makes the same point:

"Overzealous testing of easily measured objectives using multiple choice answers and the omission of any untestable processes that are essential, reduces education to training. The arts are therefore being lost, but it might be worse if the arts were to be tested.

"Scapegoating teachers and reducing them to robots who follow scripted lessons 'with fidelity' that are designed by publishing companies has killed the professionalism of talented teachers. Publishing companies are profiting most from the Common Core– one set of texts fit all–no updates needed. Profiteers selling data system management has caused spread sheets to become more important than the content of instruction."[460]

The publishing company she's talking about is Pearson, a global educational conglomerate with British educationalist Sir Michael Barber pushing Common Core across the world. Barber is committed to the green movement and says it needs to be the "ethical underpinning" of education.

In an interview with Rockefeller's Council For Foreign Relations, Barber

---

459 http://whatiscommoncore.wordpress.com/2012/07/28/usoe-the-answer-is-no-can-a-student-attend-public-school-without-being-p-20slds-tracked/
460 http://whatiscommoncore.wordpress.com/2013/09/25/data-addiction-and-common-core-dictates-destroying-schools-utah-teacher-diana-mckay-speaks-out-on-common-core/

praised the CFR for pushing global education standardisation so hard: "Can I congratulate the CFR for getting into this issue? I think it's great to see education as an issue of national security and foreign policy as well as economic and domestic policy."[461]

In a speech entitled "Whole System Revolution", Barber touches on the plan to introduce global education standards:[462]

"One of the things we see in the changing of education systems around the world is that sense of 'no more borders, no more barriers'. This is becoming a global phenomenon.

"The planet Earth can only survive until 2050 if we have a shared understanding of some of the basic ethics...all of these are key things that every child in every school around the world should be learning.

"We don't want governments to set out codes of ethics, we want teachers to think about what that means, we want schools to operate as ethical communities. That's what we want children to learn."

What was it that United Nations Assistant Secretary-General, Satanist and UNESCO founder Robert Muller said long ago had to happen in education to raise consciousness sufficient to bring back Lucifer to rule the Earth?:

"A new world morality and world ethics will thus evolve, and teachers will be able to prepare responsible citizens, workers...A world core curriculum might seem utopian today [but] by the end of the year 2000 it will be a down-to-earth, daily reality in all of the schools of the world... the agenda for our cosmic future has struck."

In 2002, New Zealand's Labour Government – then headed by Helen Clark who now leads the UN Development Programme – signed on to introduce the "Whole Child" policy approach recommended by the United Nations.

The UN Convention on the Rights of the Child (UNCROC) says at clause 2.2

"Article 2(2): States Parties shall take all appropriate measures to ensure that the child is protected against all forms of discrimination or punishment on the basis of the status, activities, expressed opinions, or beliefs of the child's parents, legal guardians, or family members."

In typical UN fashion, the clause can be read two ways. Firstly, that

461 Podcast, http://castroller.com/Podcasts/InsideCfrEvents/2695637
462 Sir Michael Barber, Chief Education Advisor , Pearson, at the Education Summit in Britain, 1 August 2012, http://www.youtube.com/watch?v=T3ErTaP8rTA&feature=youtu.be

a child should not be discriminated against on the basis of the parents' beliefs or activities. On the other hand, and because this is the UN we are talking about, the other possibility is that if the State decides something is right for the child, then interference by the parents or family to prevent that could be deemed as a discrimination or punishment of the child, justifying State intervention.

New Zealand wrapped up its UNCROC commitments in a new policy called "Agenda for Children" which incorporated the Whole Child policy at a national level, at all facets of Government. Unlike America, where people are far more suspicious of United Nations initiatives, New Zealand wears its UN dictates like a teacher's pet displaying a badge of honour. The way New Zealand officials display that obsequiousness is by clearly disclosing in their reports exactly what they think the UN wants to hear, such as this description of the State's right to redefine family structure:

"Childhood (like parenthood) is a socially constructed concept, and therefore neither universal, static nor immutable: it is what we, as a society, make it."

The NZ Agenda For Children is based on the Whole Child theories of American psychologist Urie Bronfenbrenner. These are likewise at the heart of Common Core in the US and its equivalent roll-out in the UK, Australia and elsewhere.

Like Robert Muller, Bronfenbrenner's theories are a sophisticated form of subtle indoctrination that he argued could begin at the earliest years and manifest into a "lifelong learning" meme of tweaking and re-programming even well into adulthood. Of course, no self-respecting educator ever admits they are indoctrinating, but they couch that meaning in jargon that their colleagues – ideologically wedded to the cause of shaping a new future – understand fully.

Bronfenbrenner made the point that perception can be turned into reality: "Perhaps the only proposition in social science that approaches the status of an immutable law – W I Thomas's inexorable dictum: 'If men define situations as real, they are real in their consequences'."[463]

Commenting on this, US educational analyst and former attorney Robin Eubanks writes:[464]

---

463 Which is why if the UN is committed to trying to usher in Armageddon because it wants the return of "The Coming One", you need to be worried
464 http://www.invisibleserfscollar.com/imitating-the-ussr-in-striving-to-discover-how-the-child-can-become-what-he-not-yet-is/

"False beliefs and supplied manipulative mental concepts can be, and in fact are being designed to be, hugely influential on future behavior. Real in their consequences. And we may all think of education as being about the transmission of knowledge from the past, but the social scientists think education is all about changing future behaviors. Changing moral and spiritual values and feelings and all the components of the personality that will turn into the adult's character. And the behavioral scientists, especially in the US at the premier ed [teacher training] schools, have controlled the agenda for decades."

The Lucis Trust also had big ideas on how to indoctrinate children:

"Reorient the knowledge, the consciousness aspect or the sense of awareness in the child in such a manner that he realises from infancy that all he has been taught is with the view to the good of others more than of himself."

Lucis wants children to be taught the values that encourage service to the greater spiritual and environmental plan.

"Everywhere and in every country men are taught to be exponents of certain group ideologies. These ideologies are, in the last analysis, materialising dreams, visions, or ideas. Leading educational thinkers and organisations today, including UNESCO at the world level, are increasingly emphasising the moral, ethical and spiritual objectives and needs of the educational process."[465]

The Satanic sect view sounds suspiciously close to the Agenda 21 vision as well:

"Even in countries with strong education systems, there is a need to reorient education, awareness and training so as to promote widespread public understanding, critical analysis and support for sustainable development.

"The core themes of education for sustainability include lifelong learning, interdisciplinary education, partnerships, multicultural education and empowerment."

Teachers might be surprised to learn that Agenda 21 requires them to be the puppets of the new social engineering:

"Special attention should also be paid to the training of teachers, youth leaders and other educators."

Can't have people remaining in classrooms who aren't 'in with the programme', eh what!

---

465 "Education in the New Age" Lucis Trust World Goodwill Key Concepts, http://www.lucistrust. org/en/service_activities/world_goodwill/key_concepts/education_in_the_new_age

Pearson Education, a sister company to Penguin Publishing and the *Financial Times* newspaper, is a major global player in education text books. In pushing for "global education standards" in partnership with the United Nations, Pearson stands as the gatekeeper to education ideology worldwide. Anyone else wanting the ear of education authorities will have to be singing from the same globalist hymn sheet.

Barber makes no secret of what drives him:[466]

"Given the state of the global economy, tensions in international relations, massive gaps between wealth and poverty, the deepening threat of climate change and the ubiquity of weapons of mass destruction, our contention is that we need a generation better educated, in the broadest and most profound sense of that word, than ever before."

Back in 1997, Barber caused a few feathers to ruffle when he suggested schools should replace churches as the home of ethics, starting with the concept of "global citizenship". Christianity, he argued, while "still hugely influential historically and culturally", could "no longer…claim unquestioning obedience."

Not like education administrators, then?

"For a while in the mid-20th century it seemed as if communism might establish new ethics, but by the 1970s all that remained in Western countries was rampant consumerism and 'the quicksand of cultural relativism'—an abandonment of the morality of right and wrong.

"In the absence of God and Marx what are we to do?" he posed.

The end result was the new British standard curriculum introduced in 2002 incorporated global citizenship concepts, just as the New Zealand, Australian and now American education systems are doing.

The tame NGO's of the United Nations, like British charity Oxfam, are also getting in on the act. Oxfam has produced a document helpfully entitled "Education for Global Citizenship: A Guide For Schools", which offers the usual homilies to the need to change human behaviour by indoctrinating children worldwide. High school students, for example, should be fully briefed on "the key issues of Agenda 21", and should know their "role as a Global Citizen". Lessons should begin with new entrants, and parents should be encouraged to reflect global citizen ideals as well, says the charity.[467]

---

466 http://www.universityworldnews.com/article.php?story=20130316093956321
467   http://www.oxfam.org.uk/~/media/Files/Education/Global%20Citizenship/education_for_
global_citizenship_a_guide_for_schools.ashx

The Common Core curriculum has also drawn on the UN-inspired International Baccalaureate programme. One man who knows this is US attorney Robin Eubanks. Well, he was an attorney before becoming a venture capitalist in the health industry leading to the establishment of a company eventually traded on the New York Stock Exchange.

Healthcare, says Eubanks, was the perfect training ground to learn bureaucracy speak.

"A background in Law is also excellent preparation for determining precisely what the terms commonly used actually mean," he adds. "Especially in an industry that is consciously using language to hide the actual intended goals. My experience allowed me to recognize that education in the US and globally has been, for decades, engaged in a massive Newspeak (as in George Orwell's *1984*) campaign that creates a public illusion on what is being promised and what is coming to the schools and classrooms that are this country's future.

"I know what the words and terms really mean to an Ed insider and how it differs from the common public perception. I have documented what was really behind the reading wars and math wars. I have pulled together what the real intended Common Core implementation looks like. And it is wildly different from the PR sales job used to gain adoption in most of the states."

It's actually a really good point. When you have to wade through UN/Government/Education/Health policy documentation, it's as if the stuff is written using molasses as ink. It just goops around your cognitive functions and slows you down. A similar effect arises when you read the Lucifer documents – other commentators have independently remarked feeling "physically ill" after reading Lucis Trust material.

On his blog, dedicated to analysing the Common Core deception, Eubanks writes:

"Those of you past a certain age may remember the 80s TV show "The A-Team" when George Peppard would put a cigar in his mouth, lean back, and with a grin say 'I love it when a plan comes together'. Well today we really are taking a huge step towards unravelling a well-laid but nefarious scheme that involves Common Core but more importantly it involves education globally. And UNESCO. And the IB, International Baccalaureate Program, and its IB Learner Profile and concept of Global Citizenship as where Common Core is actually going."

Eubanks explains that while schools and education experts are pushing the benefits of "national standards" in the mainstream media – and getting positive parent buy-in to the process, behind the scenes the planners behind national standards/Common Core are discussing things like how to subconsciously break the wills of students and make them conform to the new paradigm to come.

The way they're doing it is a classic Soviet communist technique; putting students in emotionally confronting situations designed to challenge their existing belief structures. First, break the student's ties to their own values, then fill the vacuum with the kool aid of choice.[468]

You might recall how Georgetown University's Professor Quigley lamented, in 1966, the trend towards teaching students to be "intuitive and subjective". Rather than teaching facts, as such, the new education curriculum is deliberately putting children in situations that would emotionally challenge adults, in order to penetrate a student's youthful defences.

"The easiest way to explain what is being sought," writes Eubanks,[469] "is a desire to have all thought grounded in emotions. It is the constant refrain that the problems to be used for assessment have no fixed answer and it is why lecturing and textbooks are becoming abhorrent. They build up the logical, independent, mind and are not necessarily grounded in feelings. Which means they may not produce the behavior desired to fit with all these plans for transformation. To get that requires a personality that has been shaped by 'qualitative metamorphoses in affective-cognitive experiencing and thinking.' Which is precisely what the new curricula and gaming and online learning and these new assessments are designed to create. It's also why you keep hearing so many mentions now globally to 'quality learning.'

---

468 A classic example of the genre is gay psychologist Dr John de Cecco's groundbreaking research on changing sexual orientation. If you listen to the liberal mainstream media, the idea of changing sexual orientation is a con trick, a conservative lie. People are born gay, right? Not according to De Cecco. In his book, "If You Seduce A Straight Person Can You Make Them Gay?" he details studies that prove the point in the affirmative, time and time again. Repeated exposure to gay porn and gay men was capable of breaking the heterosexual orientation of men and turning them exclusively gay within four years. The Jesuits boasted they only needed seven years with a boy to define the man. Schools have your kids for 12 or 13 years, and attendance is compulsory. Changing a child's mind about indigenous spiritualism and the evils of the West is much easier than changing sexual orientation. You can take it as read, the Pied Piper is tuning up his flute, and we're all about to reap the whirlwind. *If You Seduce A Straight Person Can You Make Them Gay?*, ed. By Dr. John De Cecco; New York: Harrington Park Press, 1993, pgs. 129-130

469 http://www.invisibleserfscollar.com/motivationally-misleading-situations-and-wicked-decision-problems-imposing-psychological-experiments-on-students/

"That's what these ill-defined 'motivationally misleading situations' and 'wicked decision problems' assessments force. Discomfort in the student so they change values and strategies and how they view the world. Such 're-examinations are actual executive-learning situations, where the subject, little by little, can acquire suitable meta-executives' that will guide the desired 'mental revolution' of when and how student's choose to act going forward in dealing 'with the hard, misleading reality of everyday experience'.

"That's why the problems have to be authentic and grounded in the real world and relevant. It brings in emotions and changes how the world will be viewed going forward," says Eubanks.

The Soviets would call it 're-programming'.

Some educational psychologists suggest will-power within students offers a path for kids to maintain resistance, while others go so far as to suggest the human will is an illusion in these new global circumstances.

Other cognitive psychologists and education professors, explains Eubanks "are saying no, human will exists but we can use pedagogy and theories of education to both shut it down and guide it in desired ways. Since we would all rebel if that was the way these reforms were presented, they are not being phrased that way. To us. In the materials we are supposed to use to frame our beliefs and attitudes toward education reforms. But I track down to the insider-only material that gets withdrawn from library shelves for a reason and it is quite clear. In fact, the commonly used term 'college and career ready' is clearly a play on gaining over time a progression of how students "create our conscious structuring of the intersubjective world of everyday experience" so that each student structures their vision of reality in the way desired [by the authorities].

"Over time these motivationally misleading situations and wicked-decision problems are supposed to create empathy in the student towards others and the world. To be 'developmentally sophisticated and advanced' in this vision of education, what is desired in future students is to be 'humanistically oriented or psychologically 'spiritual.' Now you know why we just kept encountering such a psychological emphasis as we explored the real Common Core implementation and why there is so much deceit and misleading definitions to so many terms. Being upfront and declaring you are seeking a personality suitable for the illicit political revolution may be true but it would make for a bad PR campaign," concludes Robin Eubanks.

And so the die is cast. National standards are here, globally, not to educate your children for their own benefit, but to create the kind of teenagers whose exam papers we read earlier, the shock troops of the future revolution.

Like all things, there is merit in some of the ideas, but that's the problem: globalisation is being sold only on its merits and not on the contents of the Trojan Horse that it is bringing. Worldwide, the United Nations, working through national education systems, is capturing children and indoctrinating them to accept the coming Regime. It is that simple. These kids will be voting in a few years, and they will vote to accept it because they've been conditioned to accept it. They will, in fact, expect nothing less, because this is what they have been promised their entire school lives.

And behind it all, an organisation with an altar to the Devil in its heart, right underneath the UN debating chamber, an altar where Pope Paul VI bent his knee in 1965 and surrendered his symbols of papal authority to the occult-believing head of the UN, U Thant.

The UN, its officials, their supporters in the massive liberal charitable foundations like the Rockefellers, the Fords, the Carnegies, people like Maurice Strong, are spiritually wedded to the idea they can bring back Lucifer to rule the earth, and that all this must be done in preparation.

You'll recall reading at the start of the climate change chapter about a scientific conference held on the eve of the Copenhagen Climate Summit in 2009. Around 2,500 delegates, many of them climate scientists and officials, attended that science conference and issued a joint declaration with a quote you were bound to forget over the pages that followed. The point needs to be stressed: *these were scientists.* Do you remember what they said?

"Ultimately these human dimensions of climate change (the cultures and worldviews of individuals and communities) will determine whether humanity eventually achieves the *great transformation* that is in sight at the beginning of the 21st century."[470]

Let's see the pedigree of that phrase.

"Forget not that our planet is not yet a sacred planet, though it is close to that *great transformation*," wrote Satanist Alice Bailey in a document published by the Lucis Trust (you knew that was coming).

"When *That* which overshadows Him during this incarnation has

---

470 http://lyceum.anu.edu.au/wp-content/blogs/3/uploads//Synthesis%20Report%20Web.pdf

wrought the needed changes through a process of transformation and transmutation, then a great Transfiguration will take place and He will take His place among those empowered to work through a sacred planet."[471]

Or then there's this:

"They are helping to build humanity's bridge to the Spiritual Kingdom. Therefore, there will be an ongoing need for this work, to help humanity rise to the opportunity which the Christ's presence among us provides. It is a service in which we can involve ourselves for the rest of our lives and know that we are helping in the *great transformation* into the New Age."[472]

At the World Economic Forum at Davos in Switzerland, beloved of occultist Maurice Strong and David Rockefeller, its theme for its 2012 summit was "The Great Transformation: Shaping New Models."[473]

Not to be left out, the United Nations published a 251 page report in 2011 entitled "The Great Green Technological Transformation".

It's a pretty good bet, given climate change's overt links with the pagan religion, that the science reference was in that vein.

Bat guano or not, if that's their real motivation, and it is, how far do you trust these people with your kids?

To see what these people are creating, you only have to watch a broadcast of the US TV programme Penn & Teller from a few years back. If they are manufacturing a generation of pre-programmed robots, surely there would be some evidence of that. Proof of the hypothesis was easily found by Penn & Teller when they attended a climate change protest by the Rainforest Action Network. Most of those present were university students, or in the 18-30 demographic. They were the product of the modern education system. They knew nothing about climate change, except the mantras they'd been taught.

"We won't have Florida in fifty years because of rising ocean levels," said one, ignoring the scientific studies indicating a sea level rise of about 15cm over that time – not enough to flood a beach let alone a state.

After interviewing many protestors and finding they were "well meaning" but utterly "ignorant...naïve" on the actual science, Penn & Teller concluded the debate was being sold to these people in emotional terms, creating a generation of "joiners" – people prepared to join causes without

471 "The Fourteen Rules For Group Initiation: Rule 13", The Lucis Trust, http://www.lucistrust.org:8081/obooks/?q=node/563
472 http://www.esoteric-philosophy.net/transmed.html
473 January 25-29 2012: World Economic Forum, Davos-Klosters, Switzerland

really understanding them, either because of peer-pressure or because it makes them feel good.

As proof of that hypothesis, one of the TV show's researchers used the same emotive language techniques to gather signatures for a petition to ban di-hydrogen monoxide. Nothing that the researcher tells them is a lie, it's all true:

"What it is, is it's a chemical that is found now in reservoirs and in lakes. Pesticides and different kinds of companies are using this – Styrofoam companies, nuclear companies – and now when we are washing our fruit and that, it's not coming out. Which of course means that it's ending up in our grocery stores, and in our babies' food and stuff like that. It causes excessive sweating, excessive urination."[474]

Hundreds, literally hundreds of people at the rally, signed the petition calling for a ban on di-hydrogen monoxide, and understandably so: it was pitched to them in emotive, direct impact terms, just as educators are now doing to school students. The protestors wanted to "make a difference" for the planet. "We are all interconnected," one said. They didn't bat an eyelid.

So what is di-hydrogen monoxide? You might better know it by its chemical symbol H_2O: it's water.

How could we reach a point where people attending large protest rallies accept without question what their perceived authority-figures tell them? That's exactly the same mentality that saw crowd adoration and acceptance of Adolf Hitler and his ideals.

Far from teaching genuine critical thinking, schools are training students to think critically of all worldviews *except the one they are re-programming children with*; that worldview is ground zero, it is sacred. Everything else must be filtered to see how it fits with the new state worldview. If it doesn't fit, discard it. That's what the new school system is really doing.

So finally, what have we learnt:

---

474 Penn & Teller: Environmental Hysteria, 2003, https://www.youtube.com/watch?v=2DX3lZ8peBU

CHAPTER 25

Into The West

" 'What about Rivendell and the Elves? Is Rivendell
safe?' 'Yes, at present, until all else is conquered. The
Elves may fear the Dark Lord, and they may fly before
him, but never again will they listen to him or serve
him…They do not fear the Ringwraiths, for those who
have dwelt in the Blessed Realm live at once in both
worlds, and against both the Seen and the Unseen
they have great power."
– J R R Tolkien, LOTR

When you began this book, I deliberately did not foreshadow the enormity
of what you were going to read. How could I? The story is unbelievable.

Yet there it now is, in black and white, with quotes, citations, names,
dates and places. This truly is "the big picture".

If you are on the lower rungs of the New Age or freemasonry, congratulations, I've just given you a diploma course in all the things plebs are not
supposed to be aware of until they are initiated by the elite.

If you've got this far, by the way, please don't presume I bear malice
towards the people I have labelled "Satanist". The label is an accurate reflec-
tion of the religious beliefs that they follow, in the words of the founders
of those strains of thought. The Theosophists were very clear that Satan
is real, Satan is the only god, and Satan is worshipped "by many names".
Those other names are distractions. If Blavatsky and Bailey saw one of

those names as prominent, I am happy to use it as the adjective as well.

I am sure that none of these people think that what they are doing is wrong. I'm sure they all treat their pets and small children with kindness. Nonetheless, they have installed a planet-wide policy of intolerance to belief systems that don't fit theirs, and they have used taxpayer money to establish a State religion that is now taught in schools all over the world, and they are very close to imposing some kind of global governance structure. All of this was done through subterfuge. I'm calling them out on it.

Likewise if you're a follower of the New Age or Buddhism, please, before being offended, simply read some of the documents I have cited – don't shoot the messenger. I'm not suggesting you are not good people with good hearts. What I am suggesting is that a group of people with a lust for power, both temporal and spiritual, have used ordinary people to further their agenda in a very undemocratic, intolerant way.

I am as in favour of protecting the environment as you are. I also treasure freedom. My freedom to follow what I believe, and yours to follow what you believe. I have no respect for State indoctrination of children.

What are the take home points?

The phrase "I've got nothing to hide" has a basic error at its very heart, which is this: *you assume that you know what 'they're' looking for.* If this book has shown you anything, it's that if you don't know what might be around the corner, how do you know you have nothing to fear?

Sweeping surveillance laws have been imposed by governments throughout the world, ostensibly to deal with the threat of terrorism. Whilst that might remain a 'noble cause' project, the infrastructure being put in place today can equally be used by those who follow tomorrow, with a different agenda – quelling dissent.

Your technology, nearly all of it, can be used to build a dossier on who you are, what you believe, what secret things you do in your private life, where you go, who you associate with, what you like to read or watch, what your daily routine is.

Upcoming laws and proposals for carbon accounts and fully computerised control of vehicles via GPS tracking will put in place an infrastructure capable of controlling your movements.

That global multinationals are swinging behind a global governance framework based on treaties and free trade agreements because they see an opportunity to make massive fortunes and entrench their power.

That global chemical and pharmaceutical giants have been regularly lying about the safety of their products, up to and including the use of false scientific data.

That an occult group dedicated to the return of a deity they privately call Satan/Lucifer and publicly refer to as Gaia/Mother Earth has, for several hundred years and through a number of so-called 'front' organisations, been evangelising through its own followers just as hard as the Christian Church has, for the spread of its beliefs around the world.

That the occult group in the 1870s re-ignited the almost extinct religion of Buddhism not as an authentic religion but as a façade for western esotericism/Satanism.

That the infiltration of the arts and academia with both New Age and Buddhist modernism was a deliberate evangelisation tool to create greater spread of the Luciferian doctrine through universities, movies, music and literature.

That the same group helped establish an altar to Lucifer in the heart of the United Nations building, directly below the General Assembly room.

That the head of the Roman Catholic Church, on 4 October 1965, prayed at the altar to Lucifer then symbolically swore the Church's allegiance to the United Nations and its Secretary-General.

That some of the biggest Pentecostal and liberal Christian churches on the planet are essentially preaching the same occult doctrine as the Luciferian group.

That the Lucis Trust's followers obtained high office in the United Nations, World Bank and other transnational agencies enabling them to manipulate international policy.

That the Lucis Trust followers oversaw the design of a global education policy designed to "raise consciousness" of nature worship and the spiritual relationship between humanity and nature, and to cause a corresponding hostility to Judeo-Christian theology among young people

That the purpose of raising consciousness was really to hasten the return of the Coming One/Satan/Lucifer through an increase in planetary "vibrations".

That those education policies are now mostly in place as national education standards in New Zealand, Australia, the UK, Canada and shortly the USA

That Agenda 21 and the Earth Charter are to become the guiding documents for national and local governments and bureaucracies in preparation for a global governance structure to be introduced

That the issue of Climate Change has been used as the primary evangelisation tool to prepare the way for acceptance of global government, which is why the actual science on the issue is secondary to maintaining momentum for a new structure.

You are being socially-engineered to accept a 'new world order' unquestioningly, even longingly. Why would you fight what you have been conditioned to desire?

When you list it like that, it's tempting to clutch for the familiar reassurance of dismissal – 'it's all just conspiracy theory' – as if somehow that will make the world normal again.

The problem is, it isn't theory, and technically it is not a conspiracy either.

A conspiracy is a deliberate plan by two or more people, and generally people have to knowingly buy into the plan. But what you've seen in this book cannot be explained by conspiracy theory parameters. It is not a conspiracy but the victory of a worldview, a zeitgeist that has swept the world over the past century as the result of a convergence of opportunities.

In the same way that being Left wing or Right wing is not "a conspiracy", nor is what you have read in this book. A better analogy is cancer. In this case a worldview sympathetic to totalitarianism, ironically, has spread rapidly to become the dominant theme of our time.

It's tempting to say this story is rife with undeclared conflicts of interest, but actually it is not. The information has been hidden in plain sight for those with the ability to see it. Al Gore's New Age spiritualism is in his books, Maurice Strong wears his on his sleeve, as did the UN's Robert Muller. David Rockefeller said in his *Memoirs* he was proud to be a card-carrying member of the one-world-government "conspiracy".

The altar to the "God of all" made of iron in the UN building has been a matter of public record for decades, as has been its links to the Lucis Trust and Lucifer Publishing Company to anyone who bothered to check.

Pope Paul VI's prayers at the Lucifer altar are a matter of public record as well, detailed in the *Boston Globe* of October 5, 1965, albeit no one in the media made the connection between the UN Meditation Room and the Luciferians at that time.

The Satanistic theology of Theosophy is clearly visible to anyone who wants to read their published writings, as is their hijacking of Buddhism.

The UN's Agenda 21, and the follow-up reports, as well as the Earth Charter, are all public documents, albeit they are written in that draining bureaucratese that the UN specialises in.

If anything, the only comment I could add is that this might only come as a shock because the world's mainstream media have been utterly blind to any of this, probably because it is their worldview as well. In the same way we don't hear the accents of our fellow countrymen, people and groups that share worldviews are blinded to any deficiencies in them.

Since 9/11, we have been conditioned to see the world as a series of crises – Christianity vs Islam, War on Terror, Left vs Right, poor vs rich. In truth, the most dangerous threat of all to democracy has been allowed to soldier on unchallenged.

Most of us are not going to be directly affected by an airliner hijacking. All of us will be, and are being, affected by what's been revealed in this book.

You'll recall early in *Totalitaria* I recited the "bread and circuses" line from ancient Roman satirist Juvenal. In my submission, that's exactly what this group of genuine hijackers has done – distract the public with entertainment or tragedies, while the real work to takeover the world continues behind the scenes.

The most dangerous people on the planet right now are not al Qa'ida, or your migrant neighbours from anywhere in the world. The most dangerous people on the planet are the ones about to flip the switch on total control of your life.

When you measure their long term agenda and planning, as revealed in these pages, against some of the crisis points of recent years, we are forced to ask how many global crises, from terrorism to mass shootings, to "citizen" rebellions in the Middle East, have actually been driven by these people.

Professor Carroll Quigley, the renowned Georgetown University historian, accused Cecil Rhodes and the 'secret' group of causing the Boer War and the first World War through their manipulation of history to try and create the conditions necessary for a world federal government. Quigley, you will recall, knew many of the key players personally and said he'd been given access to their private papers for the purposes of his research.

Given what you now know, is it wild conspiracy theory to suggest some in these groups would try and provoke war and tragedy in order to herd the public towards the sheep-pens of totalitarianism? Or is there enough evidence on the table now to at least give that idea some plausibility?

Quigley, who specialised in the evolution of civilisation, has also argued that democracy as we know it has only ever been truly available when the citizens were as well-armed as the State. In his epic one thousand page book "Weapons Systems and Political Stability", Quigley noted that American democracy reached its peak at the end of the 1800s when both citizens and government soldiers were armed with the same weapons.[475] The advent of the 20th century brought with it tanks, fighter jets and rockets, and tipped the balance of power in favour of the State and those who could manipulate the State. He examined ancient cultures and saw similar civilisational trends – Periclean Greece, for example, enjoyed true democracy for a time when all men were well-armed.[476]

Some of you will be wondering why there isn't a chapter devoted to gun control. There doesn't need to be. The argument for an armed citizenry is self-evident, both from the logical arguments of Jefferson, Quigley and others recited here, and from the subject matter in this book.

I have also previously written on the gun control issue.[477]

One of the key reasons the United Nations is trying to move on small arms ownership is to wipe out possible resistance down the track to its agenda. If you look at the Socialist International document I cited earlier in the book that called for global government by the UN, it also called for demilitarisation. Well, that's not entirely true. It called for the US, UK and other nations to progressively transfer their military budgets to the United Nations so that a UN peacemaking force would become the dominant military force on the planet.

At that stage of course, no single country could ever escape the clutches of the UN, and in fact if most countries were demilitarised and small arms had been abolished, nothing could stand in the way of tyranny.

"Conspiracy theory", I hear some muttering. Not when that's the point being made in UN briefing documents – the UN taking control of a global army.

I do disagree with Professor Quigley on one of his conclusions – that the balance of power is impossible to overcome. Take New Zealand. There are 4.5 million people, and 1.1 million private firearms. New Zealand's army and police forces combined can muster about 20,000 people. No matter how many armoured LAVs they use or helicopters they have, if the will of the people went against a New Zealand government the government would fall.

---

475 "Weapons Systems and Political Stability – a History" by Carroll Quigley, University Press of America, 1983 pp38-39
476 Ibid, p 307
477 http://www.thebriefingroom.com/archives/2007/08/the_gun_debate.html

It is the same in the USA. There are something like 250 million firearms in private ownership in the US. At the present point in time, those guns are going nowhere.

Unless, of course, the current generation of children being educated to accept the new regime vote to get rid of guns. Which brings the whole argument full circle.

We are being asked to trade our freedoms in return for "safety", whatever that means. But as Benjamin Franklin said, anyone willing to make that deal deserves neither freedom nor safety.

And that's the real point of this book: Now that you know what's down the rabbit hole, the issues that have perhaps been vaguely niggling at you in recent years should now be flashing in bright neon warning lights.

What does one do with information like this?

There are two ways of processing this. Firstly, you can strip the whole debate of its supernatural overlay on all sides and simply see it as a naked grab for power by followers of a particular religious belief (paganism).

You might consider that their efforts are a serious threat to democracy, and that they and the politicians and bureaucrats enabling this are in breach of fundamental principles governing the separation of church and state.

The fact that they are manipulating the education systems to make "believers" of all children under the guise of state-funded "education for all", is clear evidence of that breach of church and state.

I would be kidding you if I said that a community day of protest was going to be enough to stop this group, however. The policies they have introduced globally (Agenda 21, national education standards, climate regulations) are now interwoven into the legislation of virtually every western nation.

Barring a miracle where a large majority of the public wake up and say 'we've been conned', and they force their governments to drop these laws, nothing short of revolution will force a change of course – that's not my recommendation, it is simply an observation of fact.

If you are one of the 90% or so of the people on the planet who have a religious belief of some kind (clearly the majority of you), then you have some thinking to do.

Some of the aims of the totalitarian crowd are things everyone can agree with, but that is a separate issue from the use of genuine public concern about the environment to hide a Trojan Horse agenda to take over the world and create a new religion.

Just because you and I can agree that we should keep lakes and streams fresh and clean, doesn't mean we have to buy into the whole "Gaia is God and we need to worship her" package, nor does it mean that religious concepts should be taught to all of our kids in the name of "sustainability".

If we are truly tolerant, we should agree that genuine difference and diversity is part of being human, and that efforts to turn us into planetary "citizens" who all think and act like pre-programmed robots are a bad idea.

In other words, there are good, solid natural reasons for us to reject the new totalitarianism.

However, if you are religious, there are also good spiritual reasons to reject it. Most of us have pretty vague ideas about God and the meaning of life. The surveys show that while most people "believe" in a higher power, they are not sure who or what that is.

If this book has taught you anything, hopefully it is the truth of the old saying, "the Devil is in the detail". In your vague belief about a higher power, and your genuine concern for the environment, did you expect to be taken advantage of in a spiritual exercise to welcome Satan back to the planet? Is that really what you signed up for, or was that never really made clear to you at those circle groups and labyrinth sessions?

The Luciferian crowd clearly and genuinely believe they can bring back "the Coming One", and that this is foretold by "all the great prophets".

Well, if you believe in a higher power but have never bothered to define who that power is, I suggest now is the time to rapidly boost your learning curve.

It's time to choose sides. Who will you serve? The Lucifer sects? Jesus Christ? None of the above?

I can't make that choice for you, only you can. I was once an atheist, and on a long journey to Christianity I studied the occult for a year as well. I found Christianity stood up to the questions I had as a very aggressive, very sceptical journalist. I wrote about that journey, those questions and the evidence and answers I found, in *The Divinity Code*.

If you have questions, you may find some of the answers there.

The striking thing about the events in this book is that they fit biblical prophecy like a hand in a glove. The prophet Daniel's visions of the future saw four empires, the last of which all scholars agree is Rome. The Roman Empire didn't have a clean execution and disappear from the stage of history. Instead, Caesar converted to Christianity and Roman temporal power transferred to the Roman Catholic Church throughout the West.

Roman law remains the basis of western legal systems, including the United Nations and international law. So when the head of the Roman Catholic Church prayed before the altar in the UN and handed over his tokens of office and praised the UN as the only authority on the planet, Rome's HQ effectively transferred to New York, but Rome it remains.

Of the Roman kingdom, the prophet Daniel described it as a "world power that will rule the earth. It will be different from all the others."[478]

It is worth looking at the dream of Babylonian emperor Nebuchadnezzar, which Daniel had been asked to interpret.

"Your Majesty, in your vision you saw in front of you a huge and powerful statue of a man, shining brilliantly, frightening and awesome. The head of the statue was made of fine gold, its chest and arms were of silver, its belly and thighs were of bronze, its legs were of iron, and its feet were a combination of iron and clay.

"But as you watched, a rock was cut from a mountain, not by human hands. It struck the feet of iron and clay, smashing them to bits. The whole statue collapsed into a heap of iron, clay, bronze, silver and gold. The pieces were crushed as small as chaff on a threshing floor, and the wind blew them all away without a trace.

"But the rock that knocked the statue down became a great mountain that covered the whole earth."

That was Nebuchadnezzar's dream. Daniel explained that the statue represented four kingdoms. Babylon was the head of gold, "but after your kingdom comes to an end, another great kingdom, inferior to yours, will rise to take your place."

This happened. Babylon was later crushed by the Medo-Persian empire.

Daniel predicted they too would disappear: "After that kingdom has fallen, yet a third great kingdom, represented by the bronze belly and thighs, will rise to rule the world."

This was the Greek empire of Alexander the Great, which swept away the Persians.

"Following that kingdom, there will be a fourth great kingdom, as strong as iron. That kingdom will smash and crush all previous empires, just as iron smashes and crushes everything it strikes."

The legs, in the vision, had been of iron, and scholars regard this as Rome at the height of her glory, but, says Daniel, "The feet and toes you

---

478 Dan. 7:23

saw that were a combination of iron and clay show that this kingdom will be divided. Some parts of it will be as strong as iron, and others as weak as clay.

"This mixture of iron and clay also shows that these kingdoms will try to strengthen themselves by forming alliances with each other through intermarriage, but this will not succeed, just as iron and clay do not mix."

The United Nations with it six and a half ton iron altar, and the sworn allegiance of the Roman Catholic Church from 4 October 1965 onward certainly fits the prophecy with both its Roman heritage and its "alliances" with kingdoms.

However, according to the prophecies everyone is urging you to read, the end result is a foregone conclusion.

The eerie thing about the rise of Blavatsky and Bailey, and their infiltration of every religious belief and every political system is how complete and seamless it has been. In The Matrix, humanity was promised peace, love and happiness if they kept taking the blue pill. Those who did so were blissfully unaware their body energies were being harvested by parasites as they continued 'living the dream'.

The reality for those who took the red pill was to suddenly see the enormity of what had captured their minds – that what they perceived as perfectly normal actually wasn't.

If you stand back, objectively, for a moment, consider this: how much of our daily life and global attention is now devoted in some way, directly or indirectly, to complying with the policies and worldviews laid out in this book. Our urban planning laws are all Agenda 21 based. Our social interactions are increasingly having to fit an Agenda 21 formula. Our education systems are devoting whole chunks of school time to indoctrination. In our workplaces various laws force us to think in the Agenda 21 way. In the media we debate world peace on a nightly basis through the mechanism of the nightly news. It has become an obsession. The planet collectively groans as a series of crises and outrages are played out on global screens, each one of them conditioning us to react a certain way.

Like I said, there are two ways of dealing with this. You can hope that it's a natural series of coincidences and opportunities that have assisted this group to gain the control that they have, and you can fight back in a natural sense.

A good weapon is the 100 Days concept devised by New Zealand author and commentator Amy Brooke. Based on the Swiss system, Brooke's 100

Days plan would see all laws passed by a country's legislature subject to a 100 day probation period. If, during that time, enough citizens who were opposed to a new law or treaty ratification could force a referendum, then the results of that referendum would be binding on the legislature.[479]

In effect, this gives citizens the power of veto over bad laws that might sell their freedoms away.

Of course, in a situation like this where the indoctrination is so entrenched, convincing your fellow citizens that they are slow-cooking frogs may be difficult, but it's a good start.

The other way of looking at these events is supernaturally.

If you have never considered biblical prophecy up to now, then consider this.

Jesus Christ predicted events that would occur towards the end of the Age: wars, rumours of wars, earthquakes, the rise of a world governing power formed from an alliance of kingdoms. The rise of a false religion that would deceive "even the chosen, if that were possible".

Additionally, there are passages in the Bible written more than 2,500 years ago that predicted that Jews scattered around the world would return to a new Israel.

"Say to them, 'This is what the Sovereign Lord says: I will take the Israelites out of the nations where they have gone. I will gather them from all around and bring them back into their own land. I will make them one nation in the land, on the mountains of Israel. There will be one king over all of them and they will never again be two nations or be divided into two kingdoms."[480]

"I will bring back my exiled people Israel; they will rebuild the ruined cities and live in them"[481].

"I will bring you back to the land of Israel. Then you, my people, will know that I am the Lord."[482]

"Can a country be born in a day or a nation be brought forth in a moment? Yet no sooner is Zion in labor than she gives birth to her children."[483]

---

479 http://100daystodemocracy.wordpress.com/the-concept/
480 Ezekiel 37:21-22. Note that in the time of Christ Israel was split into two kingdoms. In 1948, Israel was re-created as a unified state for the first time in millennia.
481 Amos 9:14-15
482 Ezekiel 37:10-14
483 Isaiah 66:7-8. Note that Israel became a country in the space of one day in 1948 when the UN mandate over British Palestine ended and the USA instantly recognised the sovereignty of the new state of Israel.

There are other prophecies, but the main point is this – they all had to be fulfilled before the events leading to the Apocalypse could take place. Out of nowhere in 1948, in the wake of the war to end all wars that specifically tried to wipe out the Jewish race, suddenly Israel re-emerges from the sands of the desert.

Until Israel was re-created, none of the other events could happen. Now they could, now they have.

In the Book of Revelation, Jesus delivers warnings to the churches – a clear sign that the church has fallen from grace by the end. In his warning to the Church at Thyatira, he talks of a false prophetess[484] who misleads the people into immorality and worshipping false gods.

"I will repay each of you according to your deeds," he is quoted as saying. "Now I say to the rest of you in Thyatira, to you who do not hold to her teaching and have not learned Satan's so-called deep secrets (I will not impose any other burden on you): only hold on to what you have until I come. To him who overcomes and does my will to the end, I will give authority over the nations."[485]

It is interesting, given Theosophy, New Age and Buddhism's obsession with "deep secrets" of the initiates, that this should appear as a specific warning written 1900 years ago. How could the Bible know that this is precisely the issue we would be confronting today?

The odds against so many unlikely prophecies being fulfilled are high, but when you add in the emergence of a worldwide system underpinned by a religion based on Satanism, what are the odds then of this being just a random event?

And if it's not a random event, what does that mean for you, and your family?

One final thing. The act of reading this book will not bring down thunderbolts and split the earth asunder. When you close the cover, the world will look the same as it did yesterday. The sun will still be shining, the birds will still be tweeting.

However, as Tolkien wrote, not even the meadows of the Shire will be safe once all this rolls out. Just because you can't see something, doesn't mean it isn't there.

Tell your friends and family.

In the name of Jesus Christ, peace to all who have read this.

---

484 The "Gaia" belief is centred on a feminine earth spirit
485 Rev. 2:18-29

If you'd like a fully text searchable ebook version of this so you can instantly find the passage you are looking for, visit **www.ianwishart.com**

THE CRITICS ON *THE DIVINITY CODE*:

"The genius of this Kiwi author is the ability to discover those ugly facts that slay the hypotheses of scientists, philosophers, historians and novelists that God does not exist and that Jesus Christ was not a person in history but a myth. Its coverage is almost encylopedic. Wishart's skill as an investigative journalist is obvious as he takes hypothesis after hypothesis and demonstrates their inadequacy...He also has a sense of humour that lightens the concentration" – Bishop Mackey, Roman Catholic Bishop Emeritus of Auckland

"*The Divinity Code* is one of the best 'Christian' apologetic books I have read. There are a few small details that I think shows that he is not Catholic, but it is an excellent book nevertheless. Don t miss reading it if you can. – NZ Catholic

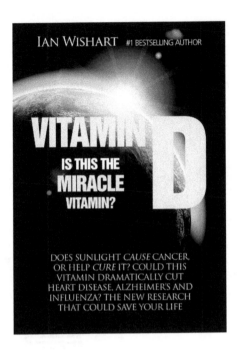

FROM THE AMAZON REVIEWS

5.0 out of 5 stars
By Dr William B. Grant, San Francisco
This book is the latest popular book on vitamin D. It covers topics of current interest including autism, cancer, erectile dysfunction, hospital-acquired infections, pregnancy, heart disease, infectious diseases, and autoimmune diseases. The research journal literature on vitamin D is growing at the rate of about 4000 papers per year yet the health system in the U.S. accepts the evidence only for falls and fractures. This book makes the case well that there are many, many beneficial effects of vitamin D. I strongly recommend this book.

5.0 out of 5 stars
By Barbara Locke
I felt this book was a great summary of the research around Vitamin D and its health effects, and the excellent referencing made it easy to read more technical research if i needed to. A very thought provoking book.

THE CRITICS ON *THE GREAT DIVIDE*:

"I can recommend it, I think it is a fascinating read and I think everybody should be reading it." – Doris Mousdale, Newstalk ZB

The Great
DIVIDE
The story of New Zealand & its Treaty

Ian Wishart #1 BESTSELLING AUTHOR

Lightning Source UK Ltd.
Milton Keynes UK
UKHW020821060123
414934UK00011B/90